The
FOCALGUIDE
to
Slide-Tape

Brian Duncalf

Focal Press • London

Focal/Hastings House • New York

BL British Library Cataloguing in Publication Data

Duncalf, Brian
    The focalguide to slide-tape.
    1. Magnetic recorders and recording   2. Slides (Photography)
    I. Title
    621.389'32   TK7881.6

    ISBN (excl. USA) 0 240 51006 2
    ISBN (USA only) 0 8038 2376 2

To my wife, Sue, without whose help
and encouragement this book would never
have been started, let alone finished.

Text set in 9 on 12 VIP Univers
Printed in Great Britain by Pitman Press Ltd.

Contents

Introduction   11

Making a Start with One Projector   14
Choice of projectors   15
A sound accompaniment   16
Tape recorders   17
The external synchroniser   20
The internal synchroniser: pulse type   20
Internal synchroniser: pause type   20
Making a start   23
Choosing music   24
Combining music and pictures   24
First run through   25

Using Two Projectors   27
A lasting effect   28
The second projector   29
Cross-fading   29
More advantages   31
Using two projectors   31
Misregister   32
Misregister: the effect of the transparency mount   32
Misregister: the effect of the projector gate   33
Projector choice   34
Quantifying the movement   35
Misregister: setting up the projectors   36
Aligning the projectors   38
Temporary misregister   40
Baseboards   40
Projector stands   41
The importance of register   42

Choice of projector lens   44
Focal length   45
Positioning of projectors   46
Stability   47
Fading   47

Methods of Cross-Fading   48
Cross-fading methods   49
Mechanical fading   49
Electric fading   52
Mechanical methods   52
Control of fading and magazine advance   54
Positioning of mechanical faders   54
Electrical/electronic methods   55
Home made electric and electronic faders   58
Commercially available electronic faders   59
Compatibility   60
Dual projectors   61

A Sound Start   62
Tape recorders and decks   63
Monophonic or stereophonic sound   63
Manual or automatic projection   64
Transportability   66
Open reel or cassette?   67
Other features to be considered   69
Tape deck or built in amplifier?   70
Pause control   70
Monitoring   72
Headphone facilities   73
Mixing facilities   73
Echo effects   74
Knobs or slider controls   74
Level metering   74
Horizontal or vertical operation   76
Compatibility with the fader   76
Amplifiers, speakers and leads   77
Amplifiers   77

Other amplifier facilities 78
Speakers 79
Impedance 80
Leads to speakers 81
Plugs and sockets 81
Microphones 82

All in Good Time 83
Memory 83
Cuesheet and stop clock 84
Use of external synchroniser 85
Use of internal synchroniser 86
Recorded verbal instructions 86
Automatic projection 87

Looking After Your Slides 88
Unmounted transparencies 89
Mounted transparencies 90
Mounting slides for dual projection 91
Mounting slides for precision in register 93
Spotting and numbering 96
Storage conditions 99

Putting a Simple Sequence Together 100
The light table 103
Editing the slides 103
Horizontal and vertical formats 106
Origination of title slides 107
Typescript 107
Transfer lettering 107
Producing title slides 108
Recording sounds 111
Editing sound 112
Copyright 117
Timing the slides 117
The initial showing and amendments 118

Getting More Ambitious 121
The object of the sequence 122
Which comes first, image or sound? 125
Visual/photographic effects in dual projection 126
Presentation/photographic effects in dual projection 132
More ambitious sound 136
Sound sources 136
Copyright 137
Live recording 137
Mixing 140
Types of mixer 141
Mixing down 141
Mixing in the commentary 142
Scriptwriting 143

Organising and Showing a Programme 149
Putting a programme together 149
Preparing to give a show 151
Packing for convenience 153
Setting up for showing 153
Possible problems 153

What Went Wrong? 155
Sequence duration 155
Overdoing the single theme 157
The subtle approach 157
The literal approach 158
Mis-matching of images and sounds 159
Regularity of changes/fades 159

Where Do We Go From Here? 161
Stereophonic sound and automatic projection 161
The four channel (quadraphonic) tape deck 162
The three channel tape deck 164
The two channel (stereo) tape deck 164
Split screen techniques 164
Three projectors on one screen 165
Automatic control 166

General considerations when using a third projector 167
Back projection 168
Stereoscopic projection 170
Multi screen 170

Appendices

1. Projector Lamps 172
2. Focusing and register slides 174
3. Image sizes for various projection distances 176
4. Devices for cross-fading two projectors 177
5. Control of more than two projectors 182
6. Cassette recorders with three and four channels 184
7. Copyright – sound recording 186
8. Copyright – photography 188
9. Slide-tape synchronisers 189
10. A circuit for a "buzz-box" 191
11. Suggested checklist 192
12. Screens and viewing conditions 193

Index 194

Acknowledgements

I would like to express my thanks to the many people who have helped me in writing this book by offering information on various specialised points. If each person were named the list would be enormous and so, for brevity, in addition to those helpful persons from companies listed in the Appendices I would like to make special mention of persons from (and these are in alphabetical order so as to avoid "preferences"!) Agfa-Gevaert, Electrosonic, Kodak, Leitz, Mechanical Copyright Protection Society, and North East Audio Ltd. In addition, many thanks to my friend, Eric Cooke, for permitting me to use some of his painstakingly created transparencies for illustrations.

Introduction

Until a few years ago an invitation to see a slide show often set one's mind in a panic trying to find suitable excuses for not becoming a willing victim to this twentieth century form of torture. Only the photographer with a reputation for good work obtained a willing and enthusiastic audience. The reasons for the former are numerous: slides were often shown at random, with no connecting theme; a series of holiday shots taken over several years, and often intermingled, necessitated the mind of the viewer to switch from a bathing beauty in Majorca to next door's dog galomphing around in last year's snow, and back to the Mediterranean again. The dialogue, equally ill prepared, completed the ordeal by being over-punctuated by "ums" and "ers" and bouts of dehydrated vocal chords.

More recently a different type of slide show has become increasingly popular. Slides shown alternately from two projectors not only avoid the blank screen in between two slides, characteristic of the single projector technique, but also offer an almost continuously changing imagery on the screen. When this is coupled with taped sound of an appropriate nature, the resulting programme has very considerable potential for competing, especially in small doses, with the more common audio visual media such as television and cinema. The production of such programmes offers considerable pleasure and satisfaction to the photographer and, in many cases, also gives a new lease of life to those who fall into the photographic rut of losing interest because "its all been photographed before". Where, at one time, a very small percentage of one's work was shown to others because only this reaches a sufficiently high, exhibition type standard, with the slide-tape medium a larger output is possible — not because the standard is made lower but because the object of the exercise is different. The aim of producing superb individual photographs is not usurped by the slide-tape medium but is simply supplemented by it. Often, the best photograph of a set cannot be used in a sequence simply because it does not "fit" a series of cross-fading images. Sometimes an aspect of photography in which one is not interested becomes a foremost objective because what a sequence is intended to say can only be achieved by becoming involved and practising such an aspect.

Some photographers maintain that the producers of slide-tape sequences are simply frustrated cine fans but this is far from true. The slide-tape medium is obviously unable to compete with cine when continuously moving images are concerned but in its own right offers advantages, especially to the amateur. Apart from the obvious advantage of lower cost and self evident superior image quality of the 35mm slide over the smaller cine format, material for a number of slide-tape sequences can be collected on the same roll of film and still not preclude the odd exhibition type shot in between. The revision and improvement of existing sequences can be achieved by simply removing a particular slide and replacing it by dropping the more recent, improved slide into the magazine. No cutting and splicing involved here! Documentary type sequences can be updated just as easily. At the time of writing much professionally produced advertising material on television is of the single-shot-dissolving-image type; no doubt economic considerations are a deciding factor here, but the slide-tape medium still has its own merits.

This book is not intended to give depth of treatment to either photographic or sound recording techniques, each of these would require a book of its own. The object is to treat those aspects of each which are of importance in the production of slide-tape sequences in sufficient depth to facilitate the readers active participation in the production of sequences which, it is hoped, will give himself and his audiences considerable enjoyment.

Throughout this book reference is made to the 35mm slide as

this format is the most common one for this medium. However, the same comments are equally valid for other formats although larger ones such as 6cm × 6cm (2¼" square) will incur considerably greater expense, in film and also more limited choice of projectors.

Finally, don't think that this medium is brand new. Audiences were enjoying cross-fading images from two "magic lanterns" during the early part of the last century!

Making a Start with One Projector

Readers who have produced sequences for two projectors are likely to skip this chapter. It does, though, mention a number of devices designed primarily for the changing of slides with a single projector, which (with a little ingenuity and modification) can provide a basis for use in more sophisticated applications.

Obviously the choice of a projector requires care. Otherwise, you can incur considerable needless expense, when your experience grows and your demands on equipment become more stringent. An automatic slide changing facility is a positive must and a remote control lead also comes high on the list of priorities. This is especially important if you expect to want non-automatic control of more than one projector. You may well already have a projector. In which case, the die is cast. Unless your finances for this sort of venture are unrestrained, the building up of projection equipment is somewhat restricted by such factors as brightness, focal length of lens and other matters discussed in the next chapter. If you are about to buy your first projector, the position at least in some respects is a happier one.

The following considerations pre-suppose that your eventual aim is to obtain two projectors for cross-fading purposes. The final choice is inevitably a personal one based on personal preferences (and, of course, financial considerations). However, it is possible that all of these factors could be over-ridden. If you are in the fortunate position of being able to borrow a projector from a friend, photo club, etc. then the purchase of an identical one gives access to two projectors for the price of one. A word of warning here, however. It is a trait of photographers to become dissatisfied with their equipment unless they feel that it is the very best which suits their purpose. So doubling up on what might be considered to be inferior equipment may, in the long term, be false economy.

Choice of projectors

Apart from the obvious factors of image quality, and refinements such as automatic focusing and so on; the most obvious way in which projectors differ is in the slide carrier and changing mechanism. Two fundamentally different designs are available, that using a straight magazine and that using a rotary magazine. Each type has its own advantages and disadvantages from the standpoint of the slide-tape programme. Straight magazines have an important cost advantage. Such magazines are considerably less expensive than the rotary type. This is an enormous advantage if you store your slides in magazines. Doubly so if each sequence is kept in its own magazine or set of magazines which, is undoubtedly the best way to work. Anyone who has had to fill up a pair of magazines with slides from boxes shortly before a showing knows how taxing this can be. Wondering whether each slide is the right way up, right way round, in the correct sequence etc. build up tensions during the show that have to be experienced to be believed. Straight magazines sold in pairs contained in their own storage boxes are particularly useful. One of the limitations of all straight magazines though, is in the comparatively low capacity; thirty-six or fifty slides is usually the limit. With long sequences of slides i.e. those with slides in excess of seventy-two or one hundred in number, slight problems arise. The same occurs when a series of sequences are to be shown at one session. A magazine change during the showing is necessary; and this can be a nerve racking experience if manual operation of two cross-fading projectors is involved. With automatic projection the situation is much easier. In either case, one magazine is presented to the projector immediately after the preceding one. The technique is a simple one and it is described, together with the relevant factors concerning magazine changes, later (see page 150).

The advantages of the free entry and exit of the straight magazine with the projector can become a disadvantage under certain extreme conditions. A colleague has told me that for showing from a balcony in a theatre he needed his projectors tilted downwards. After a slight lapse of concentration one full magazine was only just saved from being pushed out of the projector and over the edge of the balcony.

The rotary magazine can never 'escape' from the projector and can hold eighty (or in some cases up to one hundred and twenty) slides. Magazine changing is therefore less frequent. It can, though, be somewhat more awkward, especially when two projectors are used one above the other. The cost of storage of a large number of sequences in their own magazine could become prohibitive. You can store slides in much cheaper transfer storage magazines. Otherwise you are restricted to boxes. This makes the correct spotting and numbering of each slide even more important (see page 96).

A sound accompaniment

Virtually anyone who has shown slides to others has, at one time or another, added his own verbal comments. Each commentary varies according to the order of the slides and with differing audiences. Without a carefully prepared and rehearsed script, no two performances will be the same. Even the addition of a record of background music, while adding to the mood, can become distracting if the music runs out before the slides or vice versa. The only way to ensure a really polished performance is to record music, commentary and any other sound effects on tape.

The value of suitable sounds as accompaniment to a slide show is considerable. You can try a little experiment to test this. Take twenty or thirty slides on a given theme, say a collection of flower slides or perhaps a series of transparencies taken in Spain. Project these whilst simultaneously playing a record (or tape) of suitable music (eg. Beethoven's

Pastoral Symphony or some flamenco guitar music in the examples above). Change slides at a rate dictated by the music (slowly in the case of slow passages of Beethoven etc). About half way through the set of slides, switch off the sound while still continuing to project as before. The psychological effect is quite startling. The senses cry out for the music to be switched on again. This is with a casual background sound. Imagine the effect if the sound were tailored specifically to fit a series of images or vice versa. With such a programme it is often impossible to say which is the more important, the images or the sound. Each supplements the other and the programme is quite incomplete without both. This claim has been made often before and had proved to be a most controversial point. Some claim that we are, first and foremost, photographers and therefore the images must be of the greater importance.

In some cases this may be the case, but a number of the most memorable programmes I have seen consist of what appears to be, photographically, nothing better than a series of snapshots, but when shown with the appropriate sound these are transformed into something which can only be described as being hilarious. This is entertainment indeed! What satisfaction must the creators of these sequences enjoy, and such success spurs them on to even greater efforts.

Tape recorders

The only satisfactory method of adding sound to your programmes at the present time is with a tape recorder. Admittedly, for simple programmes a record could be used but for anything more ambitious the use of magnetic tape is essential. Furthermore, as we shall see later, it is possible to use the tape to record signals which will control the showing of the whole programme. (It is possible for the tape to hold information which can be used to dim the lights, draw the curtains from before the screen, show the programme and so on, but the average amateur is neither sufficiently interested in this, nor can he afford the expense involved.)

Let us consider the use of the tape recorder for the simplest of sequences employing a single projector. What functions will the recorder be required to perform?

Firstly, and obviously, it is to provide the sound accompaniment for the programmes. But do you want monophonic or stereophonic sound? This question is crucial and governs not only the choice of recorder but also other equipment that may be needed later as you become more proficient and ambitious.

The second question concerns whether you intend manual or automatic projection. Automatic projection needs a track on the tape where suitable impulses can be recorded. So, with some devices designed to change slides on a single projector, a stereo recorder can only give mono sound, the other channel being used for the control impulses. However, there is at least one device available which records impulses for a projector on the fourth track (see page 20 and Appendix 9) of the tape thereby permitting stereo sound on a quarter track recorder and automatic slide change with a single projector. (NOTE stereo sound and automatic projection using two projectors involves further complications so that a standard stereo tape recorder is unlikely to be satisfactory.)

Slide synchronisers

For the purpose of reference let us call the two types of slide-tape synchroniser *internal* and *external*. The internal type feeds an impulse onto the tape via one of the recorders built in recording heads. (On mono recorders there is only one such head and hence either sound can be recorded, or impulses for the projector – but not both. A stereo recorder is needed for the latter.) The external synchroniser has a recording head of its own and the device is placed alongside the recorder

The second question concerns whether you intend manual or automatic projection. Automatic projection needs a track on the tape on which information can be recorded for the purpose of actuating the slide changing mechanism of the

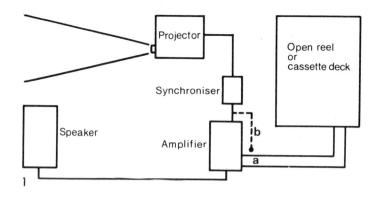

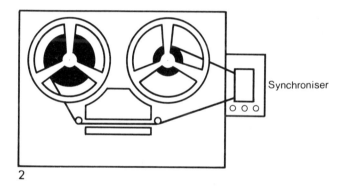

1, The internal synchroniser. a, Connection with pause type synchroniser; b, connection with pulse type synchroniser. 2, The external synchroniser. The external synchroniser is usually preferred on the right hand side of the tape recorder. Two precautions should be observed in its use. (i) For accurate synchronisation, the synchroniser must be placed in exactly the same position on playback as it was on record. The recorder may be marked for this purpose (with adhesive labels, etc.) (ii) The tape must be disengaged from the synchroniser during rapid advance or rewind of the tape.

projector via a slide-tape synchroniser. At the time of writing three fundamentally different types of synchroniser are available.

The external synchroniser

This is a self-contained unit which has its own record/playback head. It is placed at the side of the recorder in a fixed, convenient position, in such a way that the tape can pass over the head of the synchroniser either before, or after, passing over the heads of the recorder. The external synchroniser is of the quarter track type and can, therefore, be used with either quarter track or half track recorders. For obvious reasons the external synchroniser cannot be used on cassette recorders.

The advantage of the external synchroniser lies in that its record/playback head is arranged in such a position that the FOURTH track carries the impulses for projector control thereby leaving tracks 1 and 3 for stereo sound.

The internal synchroniser: pulse type

The internal synchroniser uses the recording and playback heads of the tape recorder itself. It produces a signal or pulse which is recorded on one of the tracks of the tape. On playback, the pulse is "read" and translated by the synchroniser to actuate the magazine advance switch of the projector. As the tape path is unaffected by this device, it can be used on cassette machines as well as open reel models. This advantage is not without its cost, however. As one track of the tape is needed for recording impulses, it is unavailable for sound and hence only mono sound can be used if this type of device is employed for projector control purposes.

Internal synchroniser: pause type

This is the most recent addition to the slide-tape synchroniser family and is designed to allow stereo sound to be retained.

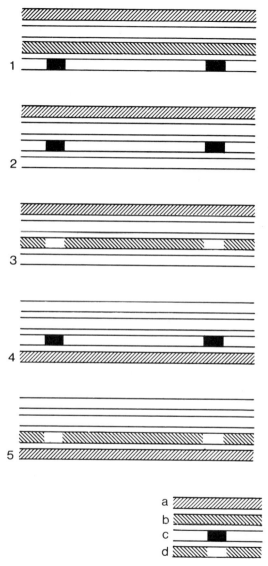

1, 2, 3 Open reels. 1, External synchroniser. 2, Internal synchroniser, Pulse type. 3, Internal synchroniser, Pause type. 4, 5 Cassettes. 4, Internal synchroniser, Pulse type. 5, Pause type. a, Left channel; b, right channel; c, Pulse; d, Pause.

The basic method of use is the same as with the pulse type but the mechanism for signalling differs. The device detects the LACK of recorded material on one of the tracks and operates the slide change mechanism accordingly. Hence a given

SOUND AND CONTROL OF ONE PROJECTOR WITH SYNCHRONISERS

Open reel		Synchroniser		Mono Sound	Stereo Sound	Projector Control
½ track	MONO	External		Yes		Yes
	(1 channel)	Internal	pulse	Yes		No
		OR	No			Yes
		pause	Yes			Yes*
	STEREO	External		Yes		Yes
	(2 channel)	Internal	pulse	Yes		Yes
			pause		Yes	Yes
¼ track	STEREO	External			Yes	Yes
	(2 channel)	Internal	pulse	Yes		Yes
			pause		Yes	Yes
Cassette						
½ track	MONO	Internal	pulse	Yes		No
	(1 channel)		OR	No		Yes
			pause	Yes		Yes*
	STEREO	Internal	pulse	Yes		Yes
	(2 channel)		pause		Yes	Yes
¼ track	STEREO	Internal	pulse	Yes		Yes
	(2 channel)		pause		Yes	Yes

* With mono sound, the pause preceding every slide change will be apparent to the audience. With stereo sound the effect will be less noticeable, being on one channel only.

slide will remain in the projector gate as long as music or other recorded sound is being read back from both channels. As soon as the recorded information on one chosen channel is interrupted for a specific duration of time (this time interval is adjustable within limits) then the slide change mechanism of the projector is operated.

Making a start

At long last we are now in a position to begin our sequence. We shall assume that, whatever equipment is available, it is possible to produce sound (mono or stereo) and also permit automatic slide change of the projector. Let us also assume that you have a collection of slides involving a single theme (say a selection of flowers), but taken over a period of time and with no specific end point in view; ie. not taken with a view to forming a sequence. Of course, you will already have projected your slides and rejected any unsuitable ones such as those which are out of focus or suffering from camera shake etc. Presumably, also, you have already decided on an order of projection.

For example, the order may conform to the general theme "gardens through the seasons" in which snowdrops are followed by daffodils, then tulips, then whatever until autumnal blooms such as dahlias are shown. If that is so, a light table (see page 103) although desirable, is not essential, as the order of slides showing is fairly straightforward.

At the outset, for simplicity, let us also assume that a commentary is not needed, just music – at least, not for the first attempt. Two considerations now arise, one being the suitability of the musical accompaniment, the other its duration. It is here that we enter the realms of personal taste and with it the biggest challenges – and also rewards. But taste apart, obviously a series of beautiful images of flowers will hardly command a sympathetic audience if accompanied by violent passages from the William Tell or 1812 Overtures (unless, of course, one intends to demonstrate how not to produce a slide-tape sequence).

Choosing music

So, for a floral sequence, you are probably looking for a piece of music of a suitably pastoral nature. Naturally, there are hundreds of such pieces, but here the second criterion appears, time. Granted, it may be that your favourite piece of music is a symphony of forty minutes duration; but this will be of little value if your sequence contains only thirty slides. Generally speaking slide changes at the rate of about eight per minute are about right, faster changes give the impression of a rushed performance and slower ones tend to bore the audience. If so, therefore, you need a piece of music about four minutes long. It is of little use to take a piece of twice this duration and chop it or fade it out half way through. Your audience will think that you haven't got enough slides (this may be true but you needn't advertise the point!). Any musical termination must be natural otherwise it will be so obvious as to jar.

It is possible to remove a middle section of a longer piece, but at this stage that is too complicated to consider. The best way is to choose a piece of appropriate duration.

Combining music and pictures

Now that you have your slides and music you can begin to put them together. At the outset we will assume that the average showing time for each slide is going to be about 7½ seconds. For goodness sake, vary the times, otherwise after a minute or two your audience will find themselves taking a deep breath and quietly (and in unison) saying "now" every time a slide change is imminent.

Arbitrary variations in time have no value either. Two factors govern the rate of change for a satisfactory performance. One is the nature of the images themselves. Images with a lot of *interesting* detail ask for a longer showing, a series of dynamic images are better suited to rapid changes. All this must, of course, link up with the second consideration: the music. For this purpose, you will have to *listen* to the music many times over. Make a note of any significant changes in chord, speed, mood, etc. Ideally, these should correspond to appropriate visual changes. A stopwatch or darkroom stop-clock can be invaluable here. After several such studies the music will have punctuated itself, almost automatically, into a series of sections. Punctuate your visual images to synchronise with the sounds and you will have the makings of an audio-visual experience.

First run through

The moment of truth has arrived: you have before you your slides, in order, in a magazine and in the projector; your taped music is ready in the recorder; in front of you are the notes for timing together with a stopwatch or stopclock; you gaze at the blank screen which stands like an artist's canvas, waiting for the magic imagery. The room is darkened, the audio equipment is switched on ready to play, you are poised to switch on the stopwatch at the appropriate moment. Switch on the projector, and instantly you realise the first mistake; either you are dazzled by a brilliant screen or the first slide is showing without any introduction.

Next time you can avoid that. Not to worry for the moment, carry on, remembering that what is needed on the screen first is either a blank slide or a title slide. This leaves time to synchronise the first image with whatever part of the music you choose.

During this first showing, keep an eye on the cue sheet on which the timing is recorded, and one eye on the stopwatch; while somehow or other watching the screen. You also need a finger poised over the projector advance button. Obviously, some assistance would be most welcome. Someone to watch the cue-sheet and stopwatch and give a signal (a tap on the arm, for example, at the appropriate time for a slide change) would help enormously. Then you can change the slides, listen to the music and concentrate on the overall effect.

After the first trial run, you may need one or two minor modifications. A slight change in order of slides, perhaps, maybe some adjustment of times of one or two of the changes. If your initial work was done thoroughly, though, you will need no major revisions.

Showing your sequence like this tends to be a bit of an ordeal. Although with practice the difficulties become progressively smaller, it would be nice to be able to sit back and watch the sequence show itself automatically. This is where the slide tape synchroniser comes in.

The synchroniser records an impulse on one track of the tape. Ideally, record it after the music has been recorded and during an actual showing of the sequence. When you replay the tape, the slide changes at each impulse. We go into some practical details of slide-tape synchronisers further on (page 85).

Now you have your tape with music and synchronising impulses recorded on it, we can sit back and watch the show.

That was all right for a first try but it could be improved. It would be better if you could avoid the blank screen during slide changes and to dissolve from one slide to the next would add a new dimension. A second projector will be needed, but the difference could be fantastic. Let's give it a try.

This is where the fun really starts!

Using Two Projectors

Why use two projectors? It involves considerably more expense, not only in more projection equipment, but in ancilliaries such as fading devices, and (if used) synchronisers. More trouble is involved in setting up the projectors and, if this were not enough, even the slides themselves need to be mounted with more care. You have to give time and attention to the order in which the slides are projected, each slide must be in its correct position and in the correct magazine. If the job is to be done properly the images and sounds must synchronise, in some cases, with an accuracy of a fraction of a second. All these things, and more, need to be kept under control, and, of course, all the more things to go wrong. Is it really worth it?

To answer the last question, one needs only to consider the increasing popularity of the dual projection slide-tape medium over the conventional single projector technique to realise that the answer is a most emphatic yes. The very fact that you are reading this now is evidence of the interest which is growing. Would you have acquired this book to read if it had been available ten years ago?

Let us not make the mistake of thinking that this medium is just a new gimmick, a fashion which people adopt and then discard when the novelty has worn off. I predict that, for the serious amateur photographer with an interest in slide-tape, the days of the single projector production are numbered; or, if not, it is probably the expense involved rather than the medium itself which inhibits his enthusiasm. To describe a dissolving-view slide-tape sequence to a person who has never seen one is an impossible task; but once he has seen such a sequence of good quality, the viewer never quite forgets it. Such is the impact of this medium, not only is it superior in every way to a show involving a single projector but it tends to have an infectious effect on many.

A lasting effect

A few years ago, for the last time as a bachelor, I was paying my annual pilgrimage in the form of a fortnight working holiday at one of the many photographic courses to which photographic fanatics subscribe. I daily contacted my fianceé and tried to explain my intention to produce a sequence of an abstract nature. The reaction was a bout of intense indifference at the idea. After returning, in an attempt to whip up some enthusiasm, I showed her all the odd-numbered slides, then the even ones (I had only one projector at the time). No change in attitude was evident. Playing the sound-track was met with "I don't like electronic music, either".

Let me explain that, at that time, her idea of photography was an occasional snap, taken as a record, when on holiday. Not to be daunted, but fearing a negative effect on my hobby, I took her to another photographic Mecca of mine. This was a festival for slide-tape. The effect was miraculous. She developed an immediate interest in how to use my camera. In the grounds of the estate where the festival was held, I found myself drumming my fingers on the nearest tree for about a quarter of an hour, waiting to borrow *my* camera.

Now, listening to a record for the first time we find ourselves assessing whether it would be usable in a sequence. I managed to afford to buy her an S.L.R. of her own so that I can now carry on with my hobby, the only diversions being constructive ones. Go on, I dare you. Try this medium, introduce your partner to it; but do so at the peril of your purse!

This is the power of this medium. You cannot understand it until you have seen it; once you have seen it you want to become involved in it; and once you are involved in it you don't want to let it go. It makes you more attentive to your surroundings, you see things which you have passed by blindly dozens of times before; you hear sounds which you blocked out previously; you become more interested in so many more ideas; you realise that, inside, you have a creative talent and that this medium can act as a vehicle for it. You need not be a great photographer or a great recorder of sounds. All you need is the idea, the rest you can work at, and produce something that will entertain not only yourself but your family, friends and wider audiences. The only limitation to what you can achieve is your own imagination.

The second projector

But enough of this. We want to know about that second projector. Now, what can two projectors do for us that a single one can not.

Firstly, and most obviously, the blank screen which a single projector gives during the change from one slide to the next is avoided. The screen, therefore, holds a continuous series of images and the attention of the audience is, similarly, continuous. This can be very valuable if you are developing a theme or making a visual statement. With the single projector the audience sees a picture, registers it, then dispenses with it as the screen becomes blank then waits for the next picture to be presented. The subject of the first image is psychologically discarded while awaiting the following one. Even with a collection of images on a single theme of say, scenic views or butterflies the viewer tends to think in terms of "will the next picture be better than the last, or not", and so on. So they see a series of discontinuous statements.

Cross-fading

With the cross-fading technique one image appears to melt away, being replaced gradually by the next. There is no specific point during this process where the viewer can state definitely that the preceding slide has switched off and the following one switched on. The result is that the screen almost literally commands attention. Even when both images are on the screen and neither predominates, a third, new picture is visible. It is constituted from elements of each of the two; but is completely different from the effect which is obtained by viewing one slide held over another. With such a sandwich, any deep shadow area obliterates any detail of the same area of the frame of the transparency placed above or below it.

For example, consider a brightly back-lit daffodil occupying the right hand half of an otherwise black transparency; and a second similar slide with the flower on the left hand side. Superimposed on each other, then each black half would obliterate the flower of the other. Now take the two slides, project one onto the screen and gradually fade it out while fading the other in. The effect is dramatically different. The eye concentrates on one half of the screen, but then the dark half shows a little detail and the attention is partly diverted.

When both images are equally bright you have a new image in its own right and the viewer tends to judge it as such. That is, the composition and overall effect is considered. There is some urgency here, for the earlier part of the image is now fading away and with it your chances of squeezing the last drop of enjoyment from the picture. Finally, the faint ghost of the first image, still there, is overpowered by the new one on the opposite side of the screen. But is it overpowered? Could it be that this is the best of all these images? Does the size and brightness of the one not balance the different size and brightness of the other to perfection? An audience will, perhaps, never know. For they are now left with a single flower again but our attention has to be led, without our being aware of it, frome one side of the screen to the other, and we've enjoyed every second of it!

Once you have seen these two slides cross-faded in this way you will have little interest in viewing them singly in future. In skilful hands this technique can be a powerful tool in a slide-tape production and more will be said about it later (see page 127). Moreover, I would go as far as to claim that any set of slides would benefit by being cross-faded in this way, whether they were originally taken with that view in mind or not.

More advantages

Other advantages also accrue from the use of a second projector and, although each will be dealt with in some depth later (see page 126), it will not be out of place to mention them here. The superimposition of images of different colours gives a new third colour which, with a little background knowledge, can be predicted. Objects, including people can be made to appear or disappear at will in a ghost-like fashion. A sense of urgency in a sequence can be developed by cross-fading in a series of jerks (rather than smoothly). This is especially effective if accompanied by appropriate "jerky" music. Images on the screen can be split or otherwise subdivided for impact or other special reasons. Attention can be changed from far objects to near ones by selective focusing with the camera followed by suitable projection. The list of such effects is quite long and one finds that new ideas emerge to solve problems which crop up when producing a sequence. It is, in fact, the sequence itself which largely dictates the fading effect required.

Using two projectors

When a single projector is used, small changes in size of image, position of image on the screen or change of format (horizontal to vertical and vice versa) have little if any adverse effect on the audience. When two projectors are employed for cross-fading these matters become of the upmost importance for there is an area of reference when cross-fading (this disappears during the slide change when one projector is used) and an audience expects that this area will be maintained. If one projector throws an image which does not superimpose perfectly over the existing one on the screen (ie. is out of register) then the effect on the audience is one of irritation. One or two such misregisters may be pardoned but an audience will almost certainly find continual misregister most disconcerting. Given the choice between seeing a series of superb images presented in a third rate fashion or a series of fairly good images presented superbly, most people would opt for the latter. Include some of the former images in the latter presentation and there is little or no doubt what the choice would be.

Misregister

What causes misregister? A number of things. They can be classified under the following headings: (i) the transparency mount (ii) the projector gate (which holds the slide) (iii) the setting of the projectors.
Let us deal with each separately.

Misregister: the effect of the transparency mount

This is most important. Cheap plastic mounts of the folding type are satisfactory for the storage of slides, for viewing with a single projector and for initial examination of transparencies for technical and aesthetic qualities, but they are quite useless when it comes to register in dual projection. Such a mount will allow the transparency to move about in it by up to a millimetre. Not much it seems, but when each slide is projected a distance of twelve feet using a lens of focal length 85mm the overall movement becomes magnified to approximately 39mm (ie. over $1\frac{1}{2}$ inches!). If successive projected images differ in position by this much the effect, especially when viewed at a distance permitted by the average living room, leaves much to be desired. If the projection distance is increased to 40 feet, using the same equipment, the error in register on the screen can be more than five inches!
It is worthy of note that the figures above relate to an individual slide in a single projector. The possible extent of misregister when using two projectors could be twice that indicated above if the transparency in one mount moved to the left and that in the other to the right.
For the majority of cross-faded images in the average sequence, slight misregister of images can be tolerated but there

are occasions when precise registration is essential for the effect required (I have seen sequences when almost all the cross-fades have needed to be in perfect register). At the outset of using the slide-tape medium, many photographers discard the thought that such details will affect the sort of sequences which they intend to produce. However, as they gain proficiency, they impose further challenges. Before long they may well develop an idea which needs precision of register. Obviously, both the mount and the mounting of transparencies must be of a satisfactory standard but we will leave further detail of this for later (see page 93).

Misregister: the effect of the projector gate

Even when transparencies are securely mounted, misregister is still possible if the position of the mount itself can move within the projector. The symptoms as seen on the screen will be different in each case. If the mounts occupy exactly the same positions in the projectors then the rectangular reference area on the screen will remain stationary during cross-fading. Unsatisfactory mounting or faulty mounts will cause misregister within this reference area, and the cure will be self evident. However, imprecise positioning of the mounts within the projectors cause the reference area on the screen to move during cross-fading. The effect is more irritating to the audience and more serious in general than the former problem. (Note that the same symptoms can be observed on the screen if mounts of different manufacture or type are mixed in the same sequence. Again, the cure is obvious).

Mount misregister within the projector is a more serious problem for two reasons. Firstly, almost every fade will advertise the fault. Remember that for two cross-fades, three slides are involved. If the middle slide is not in the exact position needed, then both fades will show a shift in image position on the screen. Hence, if only one projector of a pair is faulty in this respect then every fade will also be potentially faulty.

The second reason why this problem is serious is that it will almost invariably be beyond the control of the photographer or projectionist. The fault, if present, lies in the design and construction of the projector. Furthermore, it is only a "fault" if dual projection is involved. The photographer will have little claim on the manufacturer or supplier unless the projectors were purchased specifically for the purposes of dual projection and received the assurance that they were satisfactory for that purpose.

Another question arises here. What is "satisfactory"? A lateral movement on the screen of 3mm ($\frac{1}{8}$ inch) each way over a projection distance of twelve feet may be acceptable to one person but totally unacceptable to another. It is, though, a reasonable basis for deciding on your standards. How does the same case stand if the projection distance is increased to forty feet? Possibly it is acceptable to viewers furthest from the screen but possibly not to those close to it.

Projector choice

Generally speaking, it is unreasonable to expect that an inexpensive projector should be able to place a series of slides consistently in exactly the same position in its optical path. However, it does not follow that a considerably more expensive projector will offer proportionately more precision in slide positioning. The reason for this is not difficult to find. Slide projectors, most of the time are made and bought, to show slides in a conventional way, ie. one at a time and using a single projector. Few people notice the difference if the position of images on the screen varies by an inch or so when projected thus. Even if they did, it would matter little. As long as the images stay within the appropriate area of the screen there is little of which to complain. The only way in which such an image shift will be noticeable is by comparing the relative positions of the edge of the image and screen when projecting successive slides. With dual projection the position is quite different, as stated above.

Quantifying the movement

How is one to know whether a projector is likely to be satisfactory for dual projection in this respect? For obvious reasons we are limited to generalisations. For a slide to produce an image which is in focus and in the required position vertically and horizontally three degrees of freedom of movement must be controlled.

The first, relating to the focusing, lies in a direction along the lens axis. This is usually controlled by having a fixed stop position against which the slide is held, usually by a spring providing pressure in the appropriate direction. Almost, if not all, projectors have this facility built in as an essential feature. This provides accurate focusing of the slide, once the initial manual focusing has been made and assuming that the slide mounts are of the same thickness (automatic focusing projectors have a built-in mechanism for ensuring that this position of the slide is maintained, even though different slide mount thicknesses are involved).

The second degree of freedom of movement which needs to be restricted is that in the vertical direction. The law of gravity ensures that slides fall automatically to the lowest possible position (all praise to Isaac Newton!). So, this is unlikely to cause any problems.

The remaining, third degree of freedom, ie. that in the horizontal plane, ie. movement sideways, to left and right is the one which can cause trouble. In the case of projectors having slides fed from the side, a third stop position can be maintained, usually by some form of spring pressure. However, projectors which are gravity fed, ie. those in which the slide tray lies above the projector and the slides fall into position by virtue of their own weight, can provide problems in as much as sideways movement is not restricted to a large degree. The slide gate must be wider than the slide otherwise the latter could not fall freely into position. In the case of general purpose projectors of this type this sideways flexibility can, and does, give rise to register problems when used for dual projection. In order to prevent this, a side stop and appropriate pressure device is needed to restrict such lateral movement. In some gravity fed models which are specifically designed for dual projection work, such a device is fitted (at a correspondingly increased price!).

To summarise, side fed projectors appear, on the surface, to offer better registration than gravity fed projectors unless the latter are designed specifically for dual projection. However, there is little to choose (as far as registration is concerned) between the two types if each has been manufactured with dual projection in mind.

Finally, for readers who wish to assess the merits of different types of projectors for register, detail of a method is given in Appendix 2. The equipment needed is very simple and cheap to produce. The only other requirement is a willing photo dealer who will not object to your using his demonstration projectors for ten minutes or so!

Misregister: setting up the projectors

Probably the most obvious cause of misregister lies in the setting up of projectors. At first, it would appear that the only requirement is to put the projectors into a position such that their images coincide perfectly on the screen and the problem is solved. This, however, is impossible to achieve, at least in theory. If a perfectly rectangular image is to be formed from a perfectly rectangular object such as a masked 35mm slide, then the projection system must comply with certain conditions.

The slide and the screen must be absolutely parallel to each other. The optical axis of the lens must be perfectly perpendicular to the slide and to the screen and, ideally this optical axis must extend through the centre of the slide and the centre of the screen. Projectors are built to ensure that the optical axis of the lens passes through the centre of the slide perpendicularly but the other part of the conditions rely on the person setting up the projector.

Most of us have seen the keystone effect produced when a projector is tilted upward to ensure that the image falls centr-

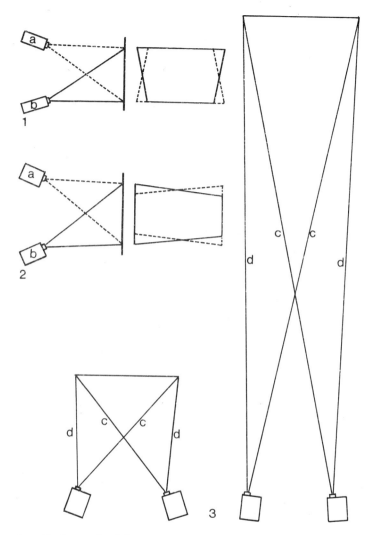

1, a, Projector tilted downwards; b, projector tilted upwards. 2, a, Level projector to left of centre of screen; b, level projector to right of centre of screen. 3, Effect of projection distance on image distortion. With a fixed projector spacing, the longer the projection distance, the smaller the distortion on the screen. For horizontally arranged projectors the ratio of the long to short vertical edges of the projected images is approximately equal to the ratio c : d.

ally on a vertical screen. A similar type of effect is obtained when tilting a camera upwards to photograph a tall parallel sided building. This effect can be avoided if the projector is placed high enough to ensure that the conditions above are complied with. Good! Now we can obtain a rectangular image with a single projector.

What happens, however, when a second projector is brought into use? Obviously, this must be placed to one side, above or below the first one and in so doing, the image from it once again deviates from the rectangular and becomes trapezium shaped. Therefore, it becomes impossible to obtain perfect register of the two images.

A compromise is needed, and the obvious practical answer is to place the projectors symmetrically on opposite sides of the imaginery line which passes through the centre of the screen and perpendicular to it. This brings us to the situation of having two trapezium shaped images if the projectors are side by side or a keystone and inverted keystone effect if the projectors are placed one above the other. However, the problem is not quite so serious as it may appear at first because these effects are most noticeable when the projection distance is short. Increasing this distance causes the effect to become progressively less obvious. Provided that the projectors are close together, it is impossible for the average viewer to observe any distortion of the rectangular image, projected over a distance of more than thirty feet or so. Once the projection distance has been increased to produce apparently rectangular images, any residual distortion will only be evident at the corners of the images. This is unlikely to cause any considerable concern.

Aligning the projectors

It is only when images of satisfactory shape have been obtained that the issue of register can be dealt with adequately. We have arrived at that point. The two images are identical as far as the eye can judge from the normal viewing distance. How can we ensure that the initial setting up of the projectors is acceptable? A number of methods are available, each with its merits and disadvantages.

(i) Two identical, blank slides can be projected onto the screen simultaneously and the projectors adjusted vertically and horizontally until the two images appear as one. The method is rather crude but offers simplicity. Accurate register may be difficult to obtain. Accurate focusing can only be achieved by focusing on the edges of the mask. The method is not entirely satisfactory.

(ii) A method similar to (i) but more refined involves using two gelatin filters of different colours in place of a transparency in the mount. Any lack of register is clearly visible as a different colour where the two images do not overlap. This method, though, is also an inadequate test for accurate focusing.

(iii) The shortcomings of the previous two methods can be serious, especially so when one considers that if the focusing of a slide is changed the *size* of the image changes. This, in turn, makes it necessary to move one of the projectors forwards and backwards until the images are exactly the same size. A modification of method (ii) involves using two identical slides consisting of a vertical and horizontal line (or series of lines parallel with the sides of the mount) in black – or much better in white (clear film), the remainder of the transparency being deeply coloured, each of a different hue. The cross lines permit accurate focusing and adjustment for the centre of the frames, and any difference in the size of the images is immediately apparent as a rim of two different colours at the perimeters of the images.

(iv) Two identical monochrome photographic negatives of a piece of graph paper, one in each projector are good for both focusing and accurate registration of images. There may be some difficulty in registering the lines at the corners but any slight misregister here which is not present at the centre of the screen is hardly likely to cause serious problems. To produce suitable negatives, the film plane and the graph paper must be perfectly parallel to each other during exposure.

(v) A modification of method (iv) involves replacing the

negative of the graph paper with that of a home-made graticule consisting of thin lines running from corner to corner and also dividing the frame horizontally and vertically (in Union flag style) (Appendix 2).

(vi) Special focusing slides are manufactured commercially for this purpose. These contain lines and geometric patterns specially designed to facilitate focusing and setting up of projectors.

Temporary misregister

Having established the causes of distortion of image shape and of misregister, and having taken steps to reduce their effects to a minimum we may consider that the troubles of this type are over. This supposition cannot be justified entirely as one final criterion must be met. The positions of the projectors must be stable; that is, neither projector must be allowed to move after the setting up process has been completed. This does not refer to the accidental nudging of one of the projectors out of the required position although this would give an obvious misregister effect. The stability refers to the base on which the projectors are placed and to the projector stand supporting this base.

Base-boards

In the excitement which comes with the acquisition of a cross-fading device, especially one of the mechanical shutter type which needs mounting on a suitable base, it is easy to obtain the first piece of plywood which comes to hand, mount the fading device and rush to observe the projected images of your slides dissolving into each other. The intention was that the base-board would be of a temporary nature only, but it is amazing how "temporary" tends to drift into permanancy when there are other demands on your time.

If the base-board is of flimsy construction (in this context 6mm $\frac{1}{4}$ inch thick plywood of the dimensions required is flimsy) then comparatively little effort will be needed to produce sufficient movement of the projector(s) to be clearly notice-able on the screen. Such movement can originate in a number of ways. Vibration of the cooling fans of the projec-tors is one possible source of continual movement although the effect is usually insignificant. The slide changing mechan-ism is a possible source of intermittent movement although in the top class projectors the action of this mechanism is very smooth.

The most likely cause of image movement on the screen is due to pressure, often by the projectionist, on one or more areas of the base-board. If the latter is insufficiently robust, a slight pressure on the corner of it (especially with a pair of heavy projectors placed nearer to the centre) will be enough to cause distortion of the base and with it, movement of one projector relative to the other. The projected image will move by a correspondingly magnified amount in spite of the great care in setting up the projectors.

A base-board of 18mm ($\frac{3}{4}$ inch) thickness should be sufficiently robust to support a pair of projectors satisfactor-ily. If, on the other hand, this base-board is intended to also support a tape deck and amplifier with associated equipment, then it may well need to be somewhat thicker. A practical test is the best and quickest way to see if a base-board is suitable. Use graticules (such as in (v) in the aligning section) pro-jected in register, simultaneously, by each projector, over the greatest practical distance. Moderate finger pressure on each of the four corners, and in other random areas, will soon establish the suitability of the base under test as the slightest distortion of the base-board will show itself in very obvious misregister on the screen. Needless to say, if the base-board is to support any other equipment under practical conditions of showing, then these items should be placed on the base, in their appropriate positions, during such a test.

Projector stands

The test above serves a second purpose of assessing the suitability of the projector stand on which the base-board is to be placed. The test is equally valid whether you use a specific projector stand for all work or whether, when showing in unfamiliar places, you have a make-shift stand (perhaps of tables).

Not only can rocking of the stand produce image movement on the screen which has an adverse effect on the audience, but it has a dreadful effect on the life expectancy of the projector bulbs. It is self evident that any instability of such a stand caused by unequal lengths of legs or uneven floors must be remedied by appropriate adjustment or packing.

The importance of register

The comments on register in previous sections may cause concern, especially with beginners in the slide-tape field. Enthusiasts with limited resources for this hobby need not be discouraged for the comments and recommendations refer to the highest standard of presentation. Many slide-tape programmes are shown using projectors from the low or middle regions of the price range and very often the audience has little of which to complain — at least as far as register is concerned.

The whole subject of register needs to be put into perspective. In many cases first attempts at slide-tape productions take the form of editing a set of slides which were taken without any consideration being given to cross-fading projection, and showing these with suitable sound accompaniment. Such sequences are quite enjoyable and encourage the photographer to use his camera in such a way as to derive full benefit from the advantages offered by the dual projection technique. Often photographers wish to go no further than this and there is no reason why they must.

However, it is a trait among photographers to try to do better, especially when they compare their efforts with those of others of more experience. It is when you become more ambitious that the subject of register arises. Still, there is no need for great concern. Misregister manifests itself in two ways.

The most common and obvious signs of misregister caused by the mount and/or projector is the movement of the entire frame of the image during the fading from one image to the next. The effect can be most disturbing to the audience. If the cause is beyond the control of the photographer, that is, if it is a projector deficiency then a partial solution is possible. The screen can be masked down with some suitable black material and the images projected onto it in such a manner (ie. taking into account projection distance and focal length of projector lens) that the edges of the image do not appear on the white portion of the screen.

In this way the edges of the image on the screen are restricted by the screen itself and not the position of the mask of the slide mount in the projector. Hence the disturbing image-area movement during fading is eliminated. The disadvantage of this method is that it restricts the showing to one specific format ie. horizontal, square or vertical. This is not a severe handicap though, as the majority of sequences are shown entirely in the horizontal form – for reasons which are discussed later. It is worthy to note that the masked screen also overcomes the keystone or trapezium effect which would otherwise be obvious when projectors are angled to the screen.

The second symptom of misregister occurs when even more sophisticated techniques of photography and showing are involved. That is, when you want to fade one part of the image into a chosen area of the preceding slide. The technique involves the use of two slides to be shown consecutively, each containing a specific area of attention of a specific size and shape, but different in content. If the shape and size is the same then a fade of considerable impact is obtained when this area is maintained but the subject changes.

Much of the impact is lost, however, if serious misregister is involved. There is no simple solution to this one. If this technique is to be employed, then projection equipment capable of giving precise registration on the screen is essential. Such techniques, though, are likely to be involved in a small fraction of the total number of sequences produced and even then, they may be used only once or twice in such a sequence.

Choice of projector lens

The projector lens is responsible for throwing a magnified image of the transparency onto the screen. As most photographers will have a camera lens of the best quality they can afford, it would be false economy to waste the detail so produced on the transparency, by projecting it using inferior optics. The projector lens should, therefore, be of a quality which will do justice to the transparencies. The remaining choice lies in the focal length. The image size on the screen depends not only on the projection distance but also on the focal length – just as it does with the camera lens. There is a difference, however, between the taking lens and the projection lens. Unlike the effect on the camera, the longer the focal length of the projector lens, the smaller the image size over the same projection distance. The advantage of the longer focal length for projection is that it allows a greater projection distance without producing inconveniently large images which would necessitate a correspondingly large screen.

Many manufacturers of projectors offer a range of lenses so that you can select a satisfactory image size on the screen within the restrictions imposed on the projection distance by the size of the room in which the projector is used. Obviously, as the projectors must be placed as close together as possible (as dealt with in a later section) the projection lenses must have the same focal lengths otherwise the image sizes would differ. Fortunately, manufacturers produce lenses of consistent focal lengths to within very close tolerances so that any small variations here will have an insignificant effect on the image size.

However, very often prospective slide-tape enthusiasts will already have a projector in his possession so that the obvious choice for a second projector will be one of the same make and specification as the first. If this were not the case then he would be ill-advised to attempt to match a projector with an 85mm lens with one having a focal length of 90mm or worse 100mm.

Focal length

If only one lens is to be obtained for each projector then there is something to be said for choosing the longest focal length which will give an image of satisfactory size in the room in which the projectors are to be used most. The benefits are twofold. The longer focal length permits the use of the projectors in rooms larger than that in which they are normally used so that sequences may not only be shown at home but also at (say) the photographic club, without resorting to very short projection distances.

The second advantage lies in that the longer projection distance will mean that, if a projector must be placed below the horizontal line passing through the centre of the screen, it will not need to be tilted upwards as much as with shorter projection distances, and hence any keystone effect will be less serious.

If you envisage much projection involving greatly differing projection distances then you need a series of lenses of appropriate focal length. However, if the range of distances involved is not so wide, then lenses with variable focal length (most of them incorrectly called "zoom" lenses) may be the answer. Most of these have variable focal length range in the order of 70–120mm.

Such lenses offer no problems if fading is to be done electrically ie. by dimming the projector lamp, but quite serious problems can arise if mechanical faders are placed in front of them. The nodal points (effectively the optical centre) of such lenses tends to lie well behind the front element of the lens. This results in vignetting of the image on the screen as the shutters close. The effect is seen as a darkening of the image from the corners or edges progressing towards the centre; instead of even dimming over the entire screen.

Positioning of projectors

The choice of the relative positions of the projectors for showing a programme would, at first thought, appear to be quite arbitrary; but a number of related factors ought to be taken into account. To a large extent the structure and functioning of the projectors themselves will have a significant effect on the choice. The two alternative positions are, of course, horizontal, ie. with the projectors placed side by side, and vertical, ie. with one projector placed above the other. Each as its own adherents, who argue its merits and offer criticism of the other. (The diagonal position, being a hybrid of the two extreme cases will be necessary if projectors having a vertically arranged rotary magazine are required to be mounted vertically and as close together as possible. As this method appears to offer few advantages while retaining the bulk of the disadvantages of each, we will make no further comment on it).

One of the first priorities is that the positioning should be such that the projection lenses are as close together as possible so as to minimise distortion of the image shape. With projectors having the slide carrier at the side, the slide changing arm may extend a substantial distance at the side of the projector during the slide change. The placing of a second projector to the right of this (looking towards the screen) will need to leave space for this action. On this count, the vertical arrangement often permits the closest proximity with convenience.

The same arguments tend to lead to the opposite arrangement for projectors with gravity feed rotary magazines. The linear spacing between lenses in the horizontal arrangement (allowing for clearance for levelling adjustments and any possible electrical connections) is in the order of fifty per cent greater than in the vertical arrangement. However, with a vertical set-up access for changing magazines tends to be awkward. This is especially important if magazines need to be changed during the showing of a sequence – remember this will need to be done in darkness!

Stability

The stability factor needs consideration. There is little difficulty when arranging projectors horizontally; but you need a suitable stand for a vertical arrangement. This stand must be open at the sides, front and back to facilitate adjustments, lead connections, lamp changing etc., and therefore support for the top projector must involve legs. For rigidity, a metal construction appears to offer the best solution.

Unless the stand is of a collapsible design (which tends to work against rigidity) its bulk constitutes a disadvantage. This is important especially if you frequently carry your equipment round.

Fading

If you are fading by dimming the lamps, the choice of vertical or horizontal arrangement can be made on other criteria; but if a mechanical method is employed, the mounting of the shutter/diaphragm adds an extra complication when projectors are mounted vertically.

The final choice of projector position is entirely in your hands. You may choose to ignore what might be regarded as a major disadvantage if you simply prefer one arrangement rather than another. In spite of everything on this subject mentioned in this book (or anywhere else for that matter) the final choice rests in the answer to two questions. Are you satisfied with what appears on the screen? Is the arrangement convenient on any criteria you wish to consider? If the answer to both these questions is "yes", then you may as well ignore criticisms and other recommendations, and carry on enjoying yourself producing and watching slide-tape sequences. After all, that is what it is all about.

Methods of Cross-Fading

Some time ago, a friend requested me to give a cine film show to the members of a society with which she was involved. When I replied that I have little interest in cine, she was quite adamant that I had given such a show the previous year. Quite frankly I was rather surprised that a slide-tape show could have been mistaken for cine but I recalled that on that previous occasion the projectors and other equipment had been housed in a separate projection room and the audience in a small theatre.

Under such circumstances a few people might be misled into thinking that such a show was of the movie type. The point here is that a well produced slide-tape performance involving cross-fading betwen two projectors can have a considerable impact and entertainment value on an audience, and bears little resemblance to the conventional slide show.

Whereas the cine show can, and sometimes does, borrow cross-fading still photographs on occasions, the reverse cannot be done without the performance becoming, quite obviously, a cine show. Cross-fading between two projectors can never produce the effect of smooth movement. In this respect cine is unquestionably superior. However, once the limitations of the slide-tape medium are appreciated then its advantages and its potential can be exploited to the full.

Let it not be thought that this medium is inferior to cine. It is simply different. Indeed the very quantity of much commercial cinema and television material which is readily available gives the slide-tape medium all the more impact by contrast. Now let us consider the factors which affect the smooth blending of one image over another on the screen. Firstly, image brightness needs to be considered. Generally speaking properly exposed colour transparencies do not vary considerably in the total amount of light which they pass (excluding areas of deep shadow and bright highlight). So the brightness of the important image areas on the screen will not vary considerably from slide to slide as long as a single projector is used.

When two projectors are used alternately their light outputs also must be reasonably similar otherwise the alternation of paler and darker images would soon become apparent. Hence we will take for granted that the light outputs are fairly well matched, even if the projectors are of different types and/or manufacture. This point concerning matched light outputs seems so self-evident that we did not think it needed to be included in the last chapter.

Cross-fading methods

Two basic methods of cross-fading are available, the mechanical and electrical/electronic. Each has its own advantages and disadvantages. Shutters, variable diaphragms or other devices can physically block out (partly or completely) the light from each of the projectors. The lamp of each projector stays at maximum brightness at all times during the performance. In the case of the electric/electronic method, the brightness of each lamp is adjusted in order to effect the fade.

Mechanical fading

In terms of cost, the former method is likely to be far lower; certainly when compared with the more sophisticated electronic methods. This method is unquestionably superior for snap changes (ie. switching almost instantaneously from one image to another). Devices for mechanical fading are not difficult to make by the average do-it-yourself enthusiast, and this may be the over-riding criterion for some.

The major disadvantage of mechanical fading is that it is not possible (at least at the time of writing) to operate the fader automatically. The presentation of a sequence or sequences must be effected by individual manual operation. Timing of rates of fade and slide changing must be achieved by memory, cue sheet or prompting as described later (see page 83). This disadvantage can be a considerable one, especially so when a lengthy programme is involved. The person at the controls is always on duty, and slight lapses of concentration can have disastrous results on the performance. For example, the missing of a slide change or changing slides twice inadvertently can send the whole performance out of phase resulting in mis-matching of visual and aural effects. Admittedly such errors can be rectified by the cool operator but they do represent potential problems which are capable of upsetting the person at the controls.

Two other problems arise with mechanical cross-fading. One involves setting up the projectors and the other vignetting. When projectors are set up for dual projection, they must be angled to each other and, if necessary, tilted to ensure that the two images are in as perfect register as is possible on the screen. This step presents little difficulty.

To obtain a smooth fading effect so that one image simply dims and the other increases in brightness evenly over the entire area of the image, it is essential that the variable aperture changes as symmetrically as possible around the optical axis of the lens. This means that the masks or diaphragms used must be very precisely placed with respect to the projectors (or vice versa if the fading devices are permanently fixed to a mounting board).

The diaphragms for fading must also be placed as close as possible to the projection lenses themselves otherwise they can cause vignetting. This effect manifests itself as an uneven dimming over the area of the projected image. With a fading image, the corners darken first and the dark areas progress towards the centre as the fade proceeds. New images start at the centre of the screen and spread outwards towards the corners. To the uninitiated, this effect may go unnoticed, but once it has been brought to your attention, the effect seems to become increasingly noticeable as time goes by. The effect can be even worse if the geometric centre of the fader diaphragm does not coincide with the optical axis of the projector

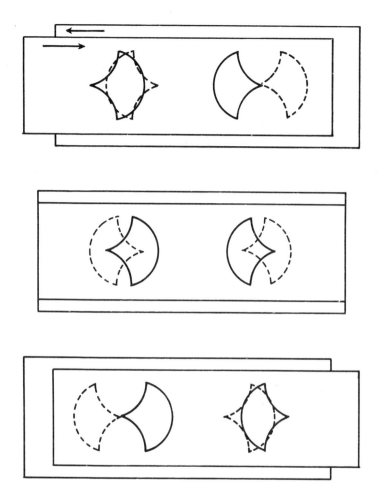

One type of mechanical fader. This requires a reciprocating action of each moving member. The device would be more versatile if the shutter in front of each lens can be opened and closed independently.

lens. Under such circumstances, the "fade" appears as a shadow creeping across the screen from one side or corner of the projected image!

The problem of vignetting is especially serious with projection lenses of short focal length, with lenses which are deeply recessed in their mounts and with vari-focal (the so-called "zoom") projection lenses. Vignetting can be minimised by careful design of the diaphragm of the fader and with care when setting up the projectors. The appearance of vignetting can also be partly obscured by staggering the opening and closing of the apertures if the faders have independent control.

Electrical fading

The electrical/electronic method of fading avoids the disadvantage of vignetting and the need to orientate mechanical devices around the optical axis of the projection lenses. Such devices can be made by those with an adequate knowledge of electronics and hence costs can be minimised.

One of the outstanding advantages of this method is that, with some commercially available equipment, the performance can be made completely automatic. This is no mean advantage when lengthy sequences involving precision in timing is concerned.

Such automation is not without its cost, however, for the fades must be programmed and such a programme must be recorded on the tape carrying the sound-track. This means that a stereo (two channel) recorder is needed and that only one channel will be available for the sound.

Mechanical methods

The ideal means of fading by mechanical means would appear to be by the use of an iris diaphragm, similar to that used in camera lenses. Of course, it must be capable of being closed completely so as to fully obliterate an image on the screen. So far, no projector lens is fitted with such a device so that the slide-tape enthusiast must rely on diaphragms placed in front. Once again, the circular iris diaphragm seems to be the best possible arrangement and if one can obtain a pair of similar diaphragms from large lenses then it would be a comparatively simple matter to construct a piece of equipment for cross-fading. However, it is unlikely that most people will have access to such material so that, for the do-it-yourself fan, the shutters themselves will have to be home made.

The simplest design is a pair of parallel blinds, the gap between which can be adjusted at will, but this can give rise to serious vignetting. An improvement can be effected by using a pair of blinds with a cut-away area giving a square aperture of variable size when operated in a reciprocating fashion. This design will give a fair fading effect with a little vignetting. A further improvement can be obtained using curved cutaway sections.

It is important to ensure that the sum of the total areas available for light to pass through *both* apertures remains constant during a fade. If it does not, the fading of the image on the screen will vary in brightness during this period.

Many home-made devices of this type have a single control arranged in such a way that when one shutter is fully open the other is fully closed. They move together so that when one is half open the other is half closed. Such an arrangement has the advantage of simplicity but lacks versatility. A better arrangement is to have independent controls for each shutter or diaphragm so as to enable each to be fully open, closed or partly open at will. The advantage of this method, if not immediately apparent, will become so in time, and with use.

The operation of the shutter mechanisms can be designed to be controlled by levers, cams or cables etc., but at the manual control end, it is of advantage to arrange that any actuating device can be operated with a single hand in such a way that at one end of the adjustment control, one projector shutter is fully open and the other fully closed, and vice versa.

This method provides simplicity of operation and retains the advantage that, you can vary the light intensity from each projector as you wish. For example, if moving both shutters simultaneously gives a vignetted effect on the screen, then you can open one shutter slightly in advance of the closing of the other shutter. So you increase the total amount of light on the screen and visually offset, at least partially, the disturbing effect of the spreading of shadows from the projector being faded out.

Control of fading and magazine advance

The best form of control is a moving lever or slider. To aid smooth operation further, you can fasten a pair of remote control leads, one for each projector, beside the controls. Position them so that the advance button for each projector lies close to the shutter closed resting point for the lever or slider. It is then simple to advance the right magazine immediately the shutter is closed.

A further refinement is possible by arranging a pair of microswitches, one at each end of the control lever or slider path so that the advance of the appropriate magazine can be affected semi-automatically. However, take care to arrange matters in such a way that you have to make a positive action to change slides. Otherwise, the magazine may advance when you don't want it to. There are occasions when advancing of magazines alternately is not wanted, for example a title slide may be required over an existing image for a short time and then withdrawn. So an attempt to make the system more automatic may cause a reduction in versatility.

Positioning of mechanical faders

As has already been stated, fading devices should ideally be within the projector lens itself. This is not possible, so the device should be placed as close as possible to the front element of the lens. Mounting shutters on the lenses themselves would appear to offer the most effective answer to this problem but the diaphragm would need to be very light in construction and the means of actuating the diaphragms would need to be suitably flexible. The shutters would have to move with the lenses, as the images were focused. This arrangement appears to offer more problems than advantages.

The arrangement of the fading devices on a permanent base-board (or boards), generally speaking, appears to be the best compromise; even though it restricts the manoeuverability of projectors. A flat board with a pair of shutters arranged perpendicularly has disadvantage of bulk and difficulty of portability. This, though, can be partly overcome by hinging the shutters so that they can be folded flat for transport. There are further complications if the projectors are to be arranged one above the other.

Finally, the construction of the shutters for mechanical fading should be such that any sliding mechanism operates in a horizontal direction. If it is arranged vertically, the law of gravity may decide to effect a fade at a different time or at a different rate to that originally intended.

Electrical/electronic methods

These methods of cross-fading have much to commend them. Operation can be manual and in some cases can offer the choice of manual or automatic. There are no physical restrictions in positioning the projectors behind, and in alignment with, the apertures of mechanical fading devices. You can choose whether to arrange the projectors side by side or one above the other. If lamp filament, mirror and condenser system are correctly adjusted (as they should be always!) then the fading of the image over the screen will be even. No vignetting is possible as with mechanical methods. In view of this impressive list of advantages, why is it that so many still prefer mechanical methods? The answer is not simply that the financial outlay of the latter is lower. They tend also to be more reliable, and can be more versatile. Every added reflnement brings with it a further function which is capable of going wrong. Mechanical devices rarely malfunction or breakdown. With electrical devices the possibility is always there, and if anything does malfunction then it is improbable that it can be rectified with sticky tape, elastic bands or string.

The versatility of the electrical systems is also, in the majority of cases, not as great as the mechanical ones. Snap changes from one projector to another are not possible, as projector lamps take a definite time to cool down and to warm up. So the fading and build up of light intensity on the screen must also require a short interval of time. (One manufacturer of an automatic electronic fading device modifies some projectors so that a mechanical shutter can be used for snap changes.) Most of the commercially available electronic fading devices work on a reciprocating system, ie. one projector can be dimmed to, say, three quarters brightness and the other builds up to one quarter brightness and so on. This means that for normal operation during a sequence both projectors cannot be operated at full brightness, nor can one be faded out without brightening the other. Admittedly you can use a blank slide at the end of a sequence; but if, for example, you use a split screen or other double image, you cannot fade out both halves of the screen together.

Naturally, such techniques are not needed for the majority of sequences, and at the outset such visual effects may be regarded as being unnecessary, but it is amazing how one's requirements become more demanding as experience and imagination grows!

One criticism of electronic fading methods is, however, not valid. Some argue that as the brightness of a bulb is dimmed, the light becomes progressively redder. This is true, but in practice such a reddening is hardly noticeable, if at all, due to the progressive superimposition of the following image. I have yet to hear a complaint that a fade was spoiled due to one of the images becoming redder!

There is one remaining point that must be borne in mind concerning the electrical or electronic method of fading. The projectors, if not designed specifically for the purpose, must be slightly modified for connection to the control system. This modification is usually quite simple and involves the

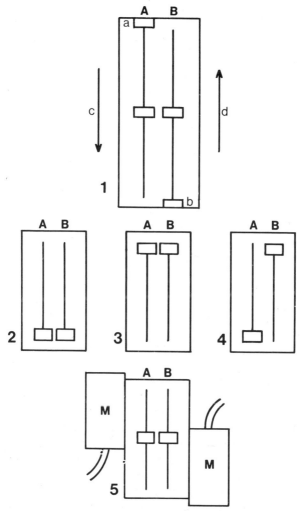

1, Suggested arrangement for hand control for electronic fading. a, Internal microswitch to advance magazine A. b, Internal micro-switch to advance magazine B. c, Brightness of projector A increasing. d, Brightness of projector B increasing. 2, 3, 4, Suggested arrangement for slider controls. 2, Projector A on, B off. 3, Projector A off, B on. 4, Projector A on, B on. 5, Suggested arrangement for slider controls and remote controls for mechanical fading of each projector. M, Magazine advance button.

connection of a pair of leads to the magazine advance switch and the breaking of one of the leads to the projector bulb for connection of the ends so produced to a socket placed somewhere on the walls of the projector. This socket provides a means of connecting the lighting circuit to the control system. To use the projector alone for conventional, non-fading projection, you need a blanking plug in the socket to remake the circuit which was broken. This offers no problems save that plugs can be a little bulky; and sometimes will not permit placing the projector in its original packing material for transport. (Hence if taking a projector to a friend's house to show some slides conventionally, don't forget the plug – bits of wire sticking out of a projector tend to be somewhat dangerous, especially in a darkened room!)

Home made electric and electronic faders

The simplest form of fading device could be a rheostat connected to the lighting circuit of each projector. (It is not advisable to vary the power to the fan of a projector as overheating could result.) One rheostat for each projector provides the versatility of infinite variety of brightness for each projector; but suffers from the disadvantage that the overall image brightness on the screen may fluctuate due to the individual controls being operated at mis-matched rates. This can be especially serious if the controls are of the rotary type. Another method involves each projector bulb being connected to the extreme ends of such a variable resistor and the slider contact being connected so as to distribute the power between the two.

The problems with simple electrical control methods outlined above is that they tend to be somewhat bulky, they are rather inefficient and produce quite a lot of heat. This necessitates adequate ventilation while, at the same time requiring effective insulation together with a satisfactory casing which is capable of preventing the operator from being accidentally connected to the mains supply!

Electronic methods of control generally involve handling

smaller power loads so that the dissipation of heat is a much less serious matter. Although the actual construction and assembly of electronic faders is not a difficult matter, the design of the circuitry is best tackled only by those with an adequate knowledge of electronics. Once again, it would be of advantage if each of the lamps could be brightened or dimmed independently. The operators slider control would have the maximum convenience and versatility if designed along the lines indicated in the figure on page 57.

The best arrangement is to have the slider controls large enough to be able to be operated individually but sufficiently close together to facilitate operation with the thumb of the hand in which it is held.

Commercially available electronic faders

A number of electronic fading devices are available in a very wide price range governed by their complexity and versatility. Some are for use only with specific projectors, others are designed for more general use. The most expensive is not necessarily the best for your purpose as individual requirements vary widely. Whereas this is not the place to discuss details of any of the faders on the market today, some general points are worth considering.

Some faders will provide for manual control only. Much thought is needed here for it can be expensive if one decides to change, at a later date, to one that will permit automatic showing of programmes.

A number of faders are available having fixed rates of fade or a choice of a limited number of rates of slide change. In my opinion, these devices are satisfactory for certain advertising purposes, but for work involving any amount of creativity they are definitely not suitable. Fixed rates of fade or fades occurring at fixed intervals soon become monotonous to the viewer and do not give the maker of slide-tape sequences the versatility which he is virtually sure to need as his skill develops. It is almost certain that, in time, you will want fades from near instantaneous to extremely slow and a similar range of slide change rates. You may want jerky changes from one slide to another and possibly also "twinkle" effects (rapidly changing from one slide to another several times). These criteria can cut down the choice of fading devices to quite a small number.

This comment refers to the use of two projectors only. When four, or more projectors are involved, a limited number of fixed fading rates facilitate the process of timing and programming. Furthermore, with multi-screen presentations, a limited number of fading rates are much less likely to be noticed by the audience.

Another factor worth bearing in mind when choosing a fading device is the ease with which you can record the control impulses for automatic showing of a programme. Some manufacturers claim that the programming can be stopped at any time and resumed later without any serious consequences. This can be no mean advantage if a lengthy sequence is involved.

Compatibility

Finally, before leaving the subject of fading devices, we must mention the subject of compatibilities. Two areas of compatibility are of importance, fader with tape deck and fader with other faders.

The fader–tape deck compatibility is best dealt with by reference to the manufacturer. Sometimes, the signal from the fader is too strong to be adjusted adequately by the input controls on the recorder. Usually, inclusion of a resistor in the lead to the tape deck is sufficient to remedy this. The reverse problem is more serious and usually the manufacturer can recommend a solution.

Fader-to-fader compatibility may appear, at first, to be of no consequence, and if you are just intending home use, this is true. However, there may be occasions where you want to show sequences at festivals, competitions etc. and this could involve transporting all your equipment (tape deck, fader and projectors). The organisers of some competitions insist on a pre-judging, usually some weeks prior to the main event. Would you be happy to send your equipment, as well as your slides and tapes, by post? In the early days of electronic faders, a control track recorded on one fader could not be relied upon to play back satisfactorily on another fader, even one of the same model type by the same manufacturer! These days, the situation is a much happier one, but it is still worth checking this point.

Dual projectors

Dual projectors are designed specifically for convenience of setting up and projection. They contain two optical systems (bulb, condenser and projection lens) built into a single unit. There is usually an in-built mechanism for varying the rate of fade and the slide change can be manual, automatic by means of an adjustable interval timer, or by means of pulses on tape.

There can be no doubt that such projectors are ideal for those who must constantly travel and use such audio-visual methods for advertising purposes or for those who wish to show their slides in a more professional manner than with a single projector, but with the minimum of fuss. However, the advantages inherent in dual projectors are offset by their limited versatility. I would suggest that the time will come (if it has not already arrived!) when you can consider an infinitely variable rate of fade and rate of slide change to be essential if you are to exploit the full potential of the slide-tape medium.

A Sound Start

The sound track for slide-tape purposes requires much attention and care if it is to serve its purpose satisfactorily. You may simply need a short piece of music or quite a complex production, made up from many different sounds from a variety of sources, all blended together. In either case, you should aim for the highest quality it is possible to achieve with the equipment available. It is worth considering why the addition of sound to a slide show (either cross-faded or not) has such a beneficial effect on the audience. Walk around some gardens admiring the flowers, for example, you may feel that the pleasure is purely visual. This is most certainly not the case. Other senses contribute; the feel of the warmth of the sunlight, the gentle touch of the breeze, the smell of the flowers – and also the *sound* of the breeze, rustling the leaves and the songs of the birds.

A simple slide show of those flowers cuts out all the other contributing pleasures, and suffers accordingly. Naturally, you cannot provide all the sensual contributions. You can, though, add sounds; and what an important contribution that is!

Imagine yourself at a crowded party where only you and your partner are talking, nobody else, no music, no other sound. Could you enjoy it? It is the variety of other sounds which makes the atmosphere. Concentrating on a specific set of sounds, however, you are not consciously listening to background. So you are not consciously aware of its presence. The sounds are there, though, and they matter. You have only to be in a room and hear a clock stop ticking to realise how much you put sounds into the backs of your minds. The cessation of the sound can have just as much impact, if not more, than the starting of a sound in the same room. There can be no doubt that a sound accompaniment is almost as important as the visual content.

Tape recorders and decks

The inadequacies of the use of records as sound accompaniment for slide shows are so self-evident that there is little point in mentioning them here. The tape recorder is ideally suited to the medium and it has no serious competitors. The choice of tape recorder needs considerable thought. As with projectors, it can be an expensive exercise to have to change it to a more suitable type later. You may already have a recorder and no intention of changing it. In which case the section which follows will serve to highlight the advantages and the limitations of that recorder.

As ambition grows and demands upon your equipment become more exacting it is likely that you will want a second tape deck; or at least access to one. If this becomes the case then it is as well to be aware of the facilities that you are likely to need. This is particularly important if you are going to buy a new tape recorder.

The questions to be considered when choosing a tape recorder include: "Do I want monophonic or stereophonic sound?", "do I intend manual or automatic projection?", "is transportability of the recorder of considerable importance?" and the ever present and all important question "how much is it going to cost?". Almost invariably the most versatile equipment is the most expensive, so that, unless money is no object, the ultimate choice involves a compromise.

At this point it will be worth considering the questions above in more detail.

Monophonic or stereophonic sound

These days pretty well all recorded music, whether on disc or tape, is intended for stereophonic replay. Most people have become so accustomed to stereophonic sound that one might be tempted to feel that only stereo is good enough for slide-tape purposes. There is no doubt that the full sound obtained by stereo is usually superior to mono even if the latter is played through two separate speakers. In addition, there are times when, to match the changing area of visual interest on the screen, a similar change in aural interest adds enormously to the effect.

There are some people who claim that sound is of secondary importance and it is the photographic image which takes precedence. With some sequences, this is undoutedly the case but with a first class, well conceived and presented sequence, it is often impossible to say whether the image or sounds are of greater importance. To use the medium to the full, neither image nor sound will stand on its own. A properly integrated sequence needs both. For example, try to imagine a documentary without the voice! If then, the sound is not of secondary importance it deserves to be treated as fully as possible. So stereo is preferable to mono.

The use of stereo sound is not without its cost however. Without considerable extra expense, stereo sound precludes automatic projector control.

Manual or automatic projection

With manual projection, you can record the sound on both channels of a stereo tape recorder. The only problem for the projectionist is that of times of fading and rates of fade. These factors will be considered in a separate chapter later.

With automatic projection, the situation is very different. Whatever control device is used, some sort of signals must be recorded on the tape. On playback, these signals operate the projectors. With a two channel (stereo) tape deck one of the channels is needed for the control signals. So the recorder is reduced to providing mono sound.

Obviously you need a third channel on the tape deck to combine automatic control with stereo sound. We are now getting into the realms of four channel (quadraphonic) tape decks and the cost of such tape decks is considerably higher than the cost of a pair of stereo machines. There is one

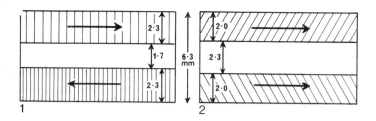

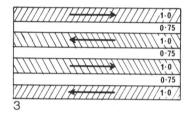

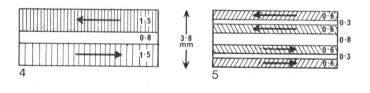

Track positions on open reel (1, 2, 3) and cassette (4, 5) tape systems.
1, Half track, mono. 2, Half track, stereo. 3, Quarter track. 4, Mono. 5,
Stereo. a, mono side A; b, mono side B; c, stereo, left channel; d,
stereo, right channel.

compromise, however. At the time of writing, there is available a cassette deck that records on two channels (with Dolby noise reduction), and a third channel reserved for control signals.

Alternatively, it may be possible to attach a separate stereo record and playback head, in an inverted position, to the existing tape recorder so that one of the heads may be used for the control signals. The electronics involved need not be buried inside the tape recorder but could be placed outside. In fact, you may be able to fit everything into the head cover if that is high enough. If you do fit an extra head, place it so that the tape can be made to by-pass it for rapid transport. Otherwise fast forward and rewind movements are likely to result in wear and tear of both head and tape; thereby reducing the effective life expectancy of both.

A further factor which needs attention if a separate head is to be mounted on the tape deck is that such a head must be mounted with precision. If the head is angled with respect to the direction of the tape travel, then the control signal put onto the tape may overlap one of the 0.75mm gaps which are intended to separate the tracks from each other. This not only means that the control impulses may be heard during the playback of quiet passages of music but also that the signals from the sound tracks may be partly superimposed onto the control impulses thereby causing faulty projector dimming and/or slide advance.

Transportability

When slide-tape shows are to be shown predominantly at the place where the equipment is normally kept (eg. home or at one's place of employment) the question of transportability is of minor importance. This factor can be enormous, however, if you show your sequences in different places and fairly frequently. It only takes a small number of occasions of transporting speakers, projectors, amplifiers, tape decks, screens and so on to make a person wonder whether he should give up this energetic pastime in favour of something like stamp collecting.

The choice here involves open reel or cassette. Naturally the latter offers less trouble in transportation and in accommodation at the site of the showing. But is the quality of sound good enough? Only the user can answer this one. Try listening to a good quality cassette recording (especially using Dolby noise reduction) and I am sure that you will not dismiss the cassette recorder lightly.

Open reel or cassette?

Neither the open reel nor the cassette system can claim superiority for use with this medium. Each has its own drawbacks. (The recently introduced Elcaset will not be discussed here because, at the time of writing, the choice of decks is limited, as is the choice of pre-recorded tapes. The latter point can be especially important if the recorder is also to serve as part of a domestic hi-fi system etc). Let us consider each of the functions required of the tape deck from the standpoint of slide-tape work.

Recording When recording the human voice, from discs, other tapes or radio in your own studio, the type of recorder matters little. Both are equally satisfactory. On-site recording, especially outdoors, is a different matter. The small size and light weight of the cassette machine win here. It is true that some portable open reel recorders of comparatively small size are available but these still tend to be rather heavy. Whether a mono or stereo machine is required is a question which, again, can be answered only by the user. The mono recorder is lighter, less costly and a stereo effect can be simulated, if needed, with a suitable mixer (see page 141).

Editing Few recordings are perfect in every respect; it may be that the sounds are in the "wrong" order; or that there are pauses; or unwanted sounds. Editing involves choosing just the right sounds in the right order, with the right pauses. This is where the open reel machine excels. In order to remove, or add a section of tape, you need precision in cutting and splicing. The precise location of the position on the tape which requires cutting and splicing is virtually impossible with a cassette as this plays back at $1\frac{7}{8}$ ips. Imagine trying to cut out an unwanted sound of a quarter of a second duration from a cassette tape! With an open reel tape recorder at $7\frac{1}{2}$ ips such a step is quite easy for a person with a little experience.

Synchronisation Internal tape synchronisation (see page 18) can be achieved with equal facility on cassette and open reel, but sound is restricted to mono if this method is used. External synchronisation, which permits stereo sound is only possible with open reel machines as, during the operation of cassette recorders, the cassette is enclosed within the recorder.

Duration of playing Cassettes are not capable of providing a continuous playback for periods of more than one hour. This is hardly a drawback, though, as the audience would not thank you for not giving them a short break during this period of time.

Showing and transportation For obvious reasons the cassette player is preferable to its open reel counterpart where transportation and playing are concerned. The compact nature of the cassette itself is also an advantage not only from the standpoint of transportation but also in avoiding the need to thread leader tapes.

Storage of tapes With open reels, one can easily collect a variety of spools in various sizes. This makes storage and indexing less convenient than with cassettes.

Choice of record/playback speeds As a general rule the quality of sound improves as the recording and playback speed increases. With cassette machines a speed of $1\frac{7}{8}$ ips is fixed (some decks have the facility of adjusting the playback speed to a limited extent so as to adjust the pitch by about a semitone, but this is largely for the purpose of ease of tuning with instruments). Tape hiss is usually quite noticeable, but this is reduced by the Dolby noise reduction system, available on many machines. With this system, hiss can be reduced to a satisfactory level.

Open reel machines usually provide two or three speeds for recording and playback. The highest speed facilitates editing as well as offering the best sound quality. Of course, then the recorder becomes more greedy in its tape consumption.

An added advantage accrues from a choice of tape speeds. Recording can be made at one speed and playback at another. A halving of speed at playback causes a drop in pitch of one octave, and vice versa. By this method Caruso can be transformed into a bass-baritone at the flick of a switch!

Mono/stereo operation It is customary on open reel stereo recorders to have separate switches for left and right channels. This means that a mono recording can be made on one channel, and at some other time a different mono recording can be made on the other. With most stereo cassette decks, the record switch operates both channels (and simultaneously erases any recording made previously). For slide-tape work this means that if one channel is to be used for timing purposes (spoken cues, aural signal or automatic projection) then the sound and timing signals must be made simultaneously. Many slide-tape enthusiasts like to run through a sequence once or twice before putting timing impulses onto tape and this can be recorded after the sound. If the cassette system is wanted for playback, but does not permit channel selection at the record stage the answer is to record the full sequence, complete with impulses (put on at a later time), on an open reel machine and copy the whole onto cassette.

Other features to be considered

Reading hi-fi magazines and manufacturers' literature on tape recorders would convince many that the ideal recorder either does not exist; or, if it does, you need a second mortgage to buy it.

This may be the case for the ardent audiophile, but slide-tape requirements are much more easily satisfied. Do you really need echo effects, foot operated remote control and simultaneous synchronisation of various sources? By and large the more sophisticated facilities built into many tape recorders can be regarded as bonus features rather than essentials. Of course, it all depends upon what sort of work you intend to do. As you become more ambitious of course, you demand more of your equipment. The following facilities are surveyed with the intention of indicating what is likely to be required for the average slide-tape fan. The choice, as always, is up to you and depends upon what sort of sound track you are likely to want to make.

Tape deck or built in amplifier?

It cannot be denied that a tape recorder with a built in power amplifier is convenient, and reduces the items you have to pack and transport. The tape deck (ie. a machine without the power amplifier) on the other hand is sometimes smaller and lighter. It is a swings and roundabouts situation. The range of recorders with a built in amplifier is smaller and the choice of power outputs is similarly limited. The extra knobs and controls for the amplifier can sometimes make the tape path to an external synchroniser a little awkward. (Remember the tape path from an external synchroniser varies considerably with the amount of tape on the nearest spool.)

The advantages of a separate power amplifier are numerous. You can choose one of suitable power output, and with a variety of other features such as filters, monitoring facilities, separate bass, treble and channel controls and so on. The range of prices is also fairly wide so that it is possible to start with a comparatively inexpensive amplifier and change to a more sophisticated model at some future date (when finances are healthier!) without the necessity of changing the tape deck at the same time. A separate amplifier also tends to be more versatile and can form the basis of a domestic (or other) sound system.

Pause control

This is sometimes called the instant stop key and is a most useful device. Its purpose is to stop movement of the tape instantly during recording or playback (but not during rapid winding) and it acts by detaching the pinch roller from the tape. This avoids the unpleasant clunk sound which would be recorded on the tape if the drive were switched off

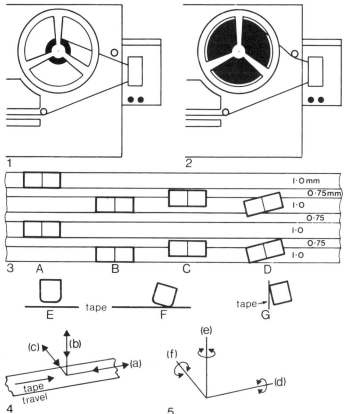

1. Tape path satisfactory at start of programme. 2. Tape chafing control knob as programme proceeds due to badly positioned synchroniser. 3. A. Normal heads; B. extra inverted head correctly adjusted with respect to (b) and (f); C. (b) maladjusted; D. (f) maladjusted; E. (c) maladjusted; F. (e) maladjusted; G. (d) maladjusted. For perfect positioning of an extra inverted head, six degrees of freedom of movement must be restricted. Three of these are movement along three directions at right angles to each other (4). 'a' is movement along a direction parallel to that of tape travel; 'b' is up and down; 'c' is towards and away from the tape. The three remaining movements are rotations (5) 'd', 'e' and 'f' around the axes 'a', 'b' and 'c' respectively. There is plenty of latitude regarding 'a' i.e. the position of the head along the tape path, but once chosen, it must be fixed. Maladjustment of any one of those remaining will probably result in unsatisfactory performance, maladjustment of two or more could cause complete failure.

electrically. It also provides a means of starting the tape with a very short acceleration time. This avoids the changing pitch which could be obtained by the drive mechanism having to start and build up to a maximum speed.

The advantage of the pause control is very evident if, for example, a voice is to be recorded intermittently over an existing sound track. If the voice is recorded on one tape, it can be transferred to the sound track at the appropriate times merely by switching the pause control to the "on" and "off" positions. More is said about this on page 142.

However, pause controls vary in operation. On some recorders, they have to be held in. These are not as convenient or useful as the switch sort, and the manufacturers do not intend them to be used for keeping the tape stationary for long periods.

Monitoring

This is the process of listening to the sounds which will be recorded onto the tape or taken off the tape. It provides a check that controls have been adjusted satisfactorily before an important recording is to be made. Obviously, the ultimate criterion of a recording, in the long term, is to listen to it over the normal recorder/amplifier-speaker system to be used but it is far more convenient to monitor during the recording so as to ensure that the quality is satisfactory. As with taking photographs you may not get a second chance!

With the less expensive open reel recorders and most cassette recorders, a single head serves for both recording and playback. This restricts us to before-the-tape monitoring, ie. listening to the sounds which are to be recorded. This gives little indication of the quality of recording or of tape hiss. The more sophisticated open reel recorders have separate recording and playback heads which permits both before-the-tape monitoring and off-the-tape monitoring, ie. listening to the sounds which have been recorded on the tape (including any tape hiss etc). Sometimes these methods are referred to as A monitoring and B monitoring respectively.

If, when choosing a recorder, all other factors are equal and the choice lies between monitoring facilities, it will be as well to choose the one with A–B monitoring.

Headphone facilities

Monitoring is very convenient using headphones. In fact, they provide the only way with live recording. Sometimes, due to a variety of distracting factors, some work on tape can be best done in the early hours of the morning. Family and neighbours can become most displeased if subjected to a large number of decibels at such times. With headphones there are no problems on that score.

With these points in mind, a socket for headphones on the tape deck must be considered to be highly desirable. Furthermore, for convenience of access this socket is best situated on the front panel or a side panel of the deck.

Mixing facilities

It is virtually inevitable that, at some time or other you will want to mix sound sources. This need is likely to increase as your experience grows and your sequences become more imaginative. When a voice is required during parts of the soundtrack this has to be superimposed over the background sound. The latter may well need to be reduced in volume during such periods.

A number of recorders are available with simple mixing facilities built in. Sometimes this takes the form of separate input controls for line (radio, record, tape) and for microphone. In other cases facilities are arranged such that line inputs are fed to one channel and microphone to another but on playback, both outputs are fed to each speaker. For individual details, manufacturers' literature or retailers of tape recorders should be consulted.

In many cases it is more convenient to tape the spoken word separately and mix this into the sound programme at a later time and for this purpose you need to be able to mix two line signals. If the tape recorder does not have any mixing facilities, or if the scope of such facilities is inadequate, then the only recourse is to use a separate mixer. (Mixers are discussed briefly on page 141.)

Echo effects

Echo effects can be obtained with most three head decks (ie. those with separate record and playback heads in addition to the erase head). The effect is obtained by using the signal from the playback head and feeding it back to the record head. As the two heads are physically separated then a time lag occurs between the two recorded sounds thereby simulating an echo effect. Obviously the reverberation time is longest at the slowest tape speed.

As echo effects are likely to be needed only on very rare occasions, such a facility in a recorder can hardly be regarded as an important factor when a choice is being made.

Knobs or slider controls

In recent years, manufacturers of recorders have tended to discard knobs in favour of slider controls. These tend to be easier in use, enabling you to pre-set controls with greater accuracy and are easier to check visually. They also tend to offer a somewhat smoother change in volume. This is especially important if one sound source is being faded out and another simultaneously brought in.

Level metering

The magic eye type of device indicates the recording level by means of two separated luminous areas on a display window. Over-recording and hence a distorted signal is indicated by the luminous areas overlapping. These devices are no longer built into recent machines as VU (volume unit) meters are now preferred. The VU meter is easier to watch and is more precise in the information given. An indication of over

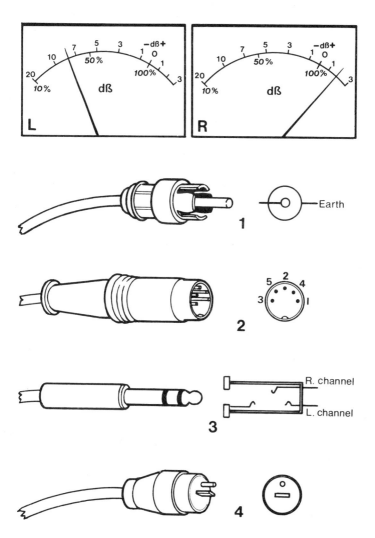

VU recording level meters. Left channel, input too low. Right channel, input too high. Plug and socket views of the following. 1, cynch plug; 2, 5 pole DIN plug (1) L input, (2) Earth, (3) L output, (4) R input, (5) R output; 3, 3-way jack plug; 4, loudspeaker plug.

recording is given by the needle moving across the scale into an area which is usually coloured red. For slide-tape work, especially with automatic projection the VU meter excels as the strength of the control signals can be monitored more accurately. This can be most important, for if the strength of the control signals is too high then it is possible to hear them during quiet passages of sound when the sequence is being shown.

Horizontal or vertical operation

The choice between horizontal or vertical operation of the recorder is largely one of personal preference and area of support needed. For vertical operation, however, the spindles for the tape reels must have suitable locking devices. (It can cause more than a little dismay, if, during the showing of a programme in a darkened room, one of the spools falls off and rolls across the floor!)
For those wishing to use an external synchroniser the horizontal position is to be preferred unless the synchroniser is firmly fixed to the side of the recorder.

Compatibility with the fader

Although this subject is the last one in this list it can be, for some, of enormous importance. For automatic projection the fading device must produce a signal which is satisfactory for the recorder. A signal which is too strong to be adjusted satisfactorily by the recorder's level controls is not a serious problem. A resistor connected in series with the lead from the fader to the recorder is all that is needed. The reverse situation is more serious. Some fading devices produce a signal which will not record at a sufficiently high level for satisfactory control when used with some tape decks. This appears to be especially the case with some cassette recorders. The only way to be quite sure on this point, apart from individual trial, is to consult the manufacturers of the fading device who should be well aware of this problem and can advise accordingly.

Amplifiers, speakers and leads

The sound produced by an audio system is much more dependent upon the speakers than upon the amplifier. So it is possible to spend a great deal of money upon an amplifier and yet produce a sound very little, if at all better than that obtained from one costing a third of the price. It is the power output and the features offered that are likely to be the deciding factors in choosing the amplifier.

Speakers are a different matter altogether. Two factors govern the choice of speakers. The first one is the pair of ears you own. No amount of advice, facts, figures or statistics can make you change your mind about whether you like a particular sound or not. Only you can choose. The other factor is the size of your bank balance or your credit worthiness.

Amplifiers

The amplifier may be required primarily as the centre of a home system but also to double up in use for slide-tape programmes. If this is the case, certain considerations may clash such as appearance, transportability, power output (a low wattage may suffice for domestic purposes but more power is needed for larger audiences of slide-tape programmes) and price. The considerations here ignore the home aspects, and concentrate on the slide-tape requirements.

Unless you have very critical hearing then it is unlikely that there will be any substantial difference in the sound produced by the large number of different amplifiers on the market. Hence the choice of amplifier will depend upon the power output and the facilities offered.

Power output If your slide-tape shows are going to be confined to the home, then the power required from the amplifier need not be large. An output of 15 watts per channel with reasonably efficient speakers in the average room should enable you to enter the ear-drum demolition business. The same equipment in a medium sized hall holding 100 or so people, may, even at maximum output, tempt the audience into feeling that they need their ears de-waxing. What I am really trying to say is that there can be no easy means of estimating the output likely to be needed. It depends very much on your speakers and the size of the room in which they operate. At a given setting an amplifier/speaker system can produce a deafening effect in an empty room with bare walls, but fill the room with people, hang curtains on the windows and fill the room with soft furnishings and much of the sound will be soaked up like water in a sponge.

If there is a general rule for helping on this subject it is always choose more power rather than less. You can always turn the volume down and it is always better to have power in reserve rather than operate at maximum output. These days, manufacturers of amplifiers tend to produce in the 20–50 watts per channel power bracket for most domestic purposes so if large audiences are anticipated, choose near the latter figure.

When comparing specifications, make sure that you compare like with like. Power ratings are normally given in either rms (root mean square) or music power. Rms figures are considerably lower for the same music power output. Also, the power output varies with speaker impedance. Typically, an amplifier rated at 50 watts into 4 ohm speakers produces 40 watts with 8 ohm units and 25 watts into 15 ohm speakers.

Other amplifier facilities

Apart from the usual facilities such as function or source switch, channel balance, tone control for treble and bass etc. some amplifiers have sockets for headphones. This facility is always useful but may be superfluous if your recorder has a similar socket. In many cases, plugging the headphones into an amplifier automatically disconnects the contacts to the speakers. For the domestic hi-fi set up this is useful for listening to records or radio, without disturbing other members of the family. Monitoring facilities of the "A–B" type are offered in some amplifiers, hence both the signal to be recorded and the signal taken from the tape can be chosen at the flick of a switch. This is a useful facility, but again, many tape decks also permit this. With A–B monitoring on the deck alone, you can hear the signal on headphones.

One very desirable feature is a "mono" selector switch. This enables a monophonic recording to be fed to both channels thereby giving a more full sound than that obtained from a single speaker. This can be quite important if automatic projection is intended (unless a 3 or 4 channel tape deck is used).

Filters are devices for reducing unwanted sounds. High filters will reduce hiss from tape and scratching sounds from imperfect records. Low filters will minimise the rumble produced by some turntables and also mains hum to some degree. Filters tend to be the sort of things that you don't bother about, until that occasion, that important occasion, when you thank your lucky stars that they are there!

Speakers

These are the one entity in a sound sytem where ears score over everything else. Paying more money does not necessarily ensure a better sound. I have compared sounds produced by speakers from different, well-known manufacturers of repute and in some cases preferred the sound produced from those costing half that of another pair of similar size. Generally though the better sound response, especially in the bass, tends to come from the speakers in the larger enclosures (and, of course, these are less transportable). Probably the best method of choosing speakers is first to establish the price bracket which you can afford (or if you are like me, cannot afford) and then compare the sounds produced. Retailers of audio equipment are usually most obliging about this. Listen to a variety of sounds such as piano, organ, brass band, full orchestra, the solo voice, choir and a speaking voice. Pop music, whether you like it or not, is not a particularly good criterion for this purpose.

Power rating Check also on the power rating which the speakers can handle. This should be in the same order of magnitude as the amplifier output. If the latter is considerably greater than that which the speakers can handle, they can be damaged if the amplifier is turned up too high during loud passages. Manufacturers are not very consistent about the way in which they rate their speakers. Some quote continuous ratings (sometimes quoted as DIN) which refers to the maximum steady rms (root mean square) power output from the amplifier which can be handled by the speakers without distortion of the sound or damage to the speaker. Another method is to quote peak ratings (sometimes referred to as music power) which is the maximum power, such as that during loud crashes of sound over a short duration, which can be safely handled. As such peaks only occur occasionally, peak ratings are higher than continuous, sometimes by as much as a factor of two.

Sensitivity For a given power output from the amplifier, the volume of sound obtained from speakers depends upon their sensitivity. The greater the sensitivity, the greater the volume of sound obtained. Unfortunately, sensitivities are not normally quoted by manufacturers so that the only indication of this is to compare the volumes obtained from different speakers when driven by the same amplifier at the same volume setting.

Over recent years, manufacturers have concentrated on producing smaller speaker cabinets without too much sacrifice in performance. The sound from some of these modern "high-pressure" units is very good, but their sensitivity is much lower. That is why the power of modern amplifiers is so much greater than was common only a few years ago.

Impedance

The impedance of speakers generally falls within the range of 4 to 16 ohms. This should match the figure quoted by the amplifier manufacturer. Too high an impedance compared to that for which the amplifier was designed will almost invariably result in a loss of volume. Too low an impedance may damage the amplifier. This can happen if two or more speakers are connected to each output channel.

Leads to speakers

Amplifier outputs are polarised, that is, they are intended to be connected in a specific way to the speakers. If the leads are connected the wrong way round, then an inferior sound will result. Amplifier outputs and speaker inputs are usually labelled positive and negative to facilitate connections. If the speakers are not so labelled, then the left hand terminal is usually the negative one. It is as well if the leads from the amplifier to the speakers are labelled in some way to assist identification, that is, if they are not colour coded already. The importance of this is not difficult to appreciate if equipment is transported fairly frequently to different places for the purpose of shows, especially if the leads are lengthy.

Finally, it is possible to pick up interference on speaker leads themselves if they are unduly long. Some manufacturers recommend that for leads over about 10 feet in length, which is most often the case when showing sequences outside the home, screened cable should be used.

Plugs and sockets

For the purpose of connecting record players, radios and tape recorders to amplifiers suitable screened leads are required. The terminals on such leads may be of two types (a) The European DIN plug which provides connections for both recording and playback on both channels using a single multi-component lead. (b) Phono (sometimes called cynch) plugs which tend to be favoured by the Japanese manufacturers. These provide connections for one type (recording or playback) signal for one channel only. They are usually sold as a pair of leads and are often colour coded for ease of identification of each channel. I have found it convenient to have sets of leads with insulation of different colours (say grey and black) so as to avoid confusion when a multitude of leads are simultaneously connected to tape recorders and amplifiers.

Microphones

The choice of a microphone is governed initially by its impedance (that is, if we exclude price as a criterion). They fall into low, medium and high impedance brackets and it is important to match microphone impedance with that of the recorder. Incorrect matching produces inferior sound quality. If a mixer is to be used this may require microphones with a different impedance to those for the recorder.

The next factor to be considered is the directional qualities of the microphone. The omnidirectional type records sounds equally well from all directions. Cardioid microphones are most sensitive to sounds coming from the direction in which they are pointed. This type is useful when you want to avoid sounds from behind the microphone.

Special-purpose microphones, too, can be useful. The Lavalier microphone is worn by the speaker (usually held by a cord around the neck), a lapel microphone is worn as its name suggests, and is useful for surreptitiously recording atmosphere sounds such as in crowds etc. When better signal-to-noise ratios are required for recording distant sounds such as a birdsong, a paraboloidal (often mis-called a parabolic) reflector can be used with the microphone or, alternatively, a highly directional tubular "gun" microphone may be employed. Both of these are bulky and somewhat expensive (the mere sight of them is sometimes enough to frighten wildlife away!).

All in Good Time

For the projectionist of a slide-tape programme there are many occasions when the timing for fades and slide changes is critical. Audiences have little care for excuses for a show which is obviously ill-presented.

Imagine a darkened room, a rapt audience watching beautiful dissolving images accompanied by suitably beautiful sounds, you are in the cockpit, manipulating the controls. To your horror, the last slide appears on the screen, and there are two minutes of those beautiful sounds still to go! Two minutes looking at one slide can seem like an eternity not only to the audience but even more so to the man in charge of operations. What do you do, let them suffer, prematurely fade the sound out, garble your apologies and simultaneously reach out for the bottle of tranquillizers?

None of these; you ensure that such a situation does not happen in the first place. It is all a matter of precision in timing.

Once the order of slides, their duration on the screen and the rate of fade has been established it is necessary to ensure that these requirements are fulfilled. How is this to be achieved? Several methods are available, depending upon the complexity of the showing and timing, and the resources available. Each of the following methods relating to manually operated fading and slide change apply equally well to mechanical or electric/electronic fading techniques.

Memory

This method relies on the projectionist being familiar with the sequence of slides and being prompted by the sound track so that certain key slides appear on the screen to synchronise with certain aural characteristics. It is satisfactory as long as a series of rapid slide changes is not needed to be in time with the sound or, for example, a slide with special impact to coincide with a sudden crashing chord. Providing that such rigid constraints are not imposed by the nature of the sequence then any error in timing at one section can be rectified later by slowing down or speeding up the rate of showing of subsequent slides.

Inevitably, a sequence will be produced which will be more rigid in its timing requirements and more precision will be needed.

Cuesheet and stopclock

There is less risk of mis-timing with this method than with the previous one. Familiarity is not so important and, indeed, even a competent stranger to the sequence should be able to project it satisfactorily if given a cuesheet.

The time for each slide is recorded on the cuesheet, ideally in two columns, one for each projector. And key slides can be marked on the cuesheet as an extra precaution.

As soon as the first sound of the track is heard a stopclock is started (a stopwatch is quieter and is likely to be more accurate, but is more difficult to keep an eye on under projection conditions). From that point on, slides are changed according to instructions on the cuesheet. You can devise suitable symbols for fades of long and short duration.

The audience may be happy with this method but you may not, as you may never be able to see your own complete sequence without the distraction of having to keep an eye on the cuesheet, and the other on the stopclock and occasionally sneaking a quick peep at what is happening on the screen. (Three independently swivelling eyeballs would be a definite asset here!)

A refinement of this technique is to employ an assistant to keep an eye on the cuesheet and stopclock or stopwatch and give you a tap on the arm, shoulder, or whatever part of the anatomy you may choose, when a fade is imminent.

Use of external synchroniser

Each of the methods described above has the advantage of enabling both channels of a stereo recorder to be used to give stereo sound. The use of an external synchroniser retains this advantage whilst overcoming the other inherent disadvantages. The external synchroniser cannot, however, be used on cassette recorders.

The standard stereo recorder uses tracks 1 and 3 for the sound, and the external synchroniser employs the fourth track for recording and playback of the control impulses. Track 2 is unused. Naturally, once track 4 has had impulses recorded upon it, the tape cannot be reversed to record another stereo programme.

The external synchroniser was designed originally for use with a single projector, the recorded impulses being used to change slides at the required time. On playback the recorded impulses are "read" by the synchroniser head and this serves to close the contacts of a microswitch, which, if connected to the projector in the conventional way, operates the advance mechanism producing a slide change. If you dispense with the connections to the projector, the lead from the synchroniser can be employed to act as a tape-operated switching mechanism for any device to which you wish to connect it. Two possibilities of use are immediately obvious. These are to provide signals for slide change which are either visual or aural.

Visual signals. A small bulb and appropriate power supply can be connected in series with the synchroniser. When a signal on the tape is picked up by the synchroniser, the contacts close and the bulb lights thereby giving a visual indication for the projectionist to commence a fade.

Aural signals. In place of the bulb above, a multi-vibrator or oscillator (which could be designed and made by any competent electronics fan) can be used to produce a signal which can be fed to the projectionist by means of a small earphone such as those used with small portable transistor radios. Such a buzz-box can be fixed permanently to the projector base-board if required.

The buzz-box should be designed to give a fairly loud sound (and preferably have a volume control) otherwise, during loud passages of music the buzz may be inaudible and the projectionist then misses a cue.

Use of internal synchronisers

Internal synchronisers act in the same way as those in the previous section and have the advantage that they can be used both on open reel and cassette stereo recorders. However, for reasons stated previously, the pulse type restricts the sound accompaniment to monophonic.

The internal synchroniser can be used to operate a buzz-box or to give visual signals, as described in the previous section.

Recorded verbal instructions

In order to avoid buzz-boxes and similar devices, there is no reason why one channel cannot be reserved to record verbal instructions. This reduces the sound accompaniment to mono, but has the advantage of being able to offer precise instructions to the projectionist. The duration of fade, twinkle effects, snap changes and even a description of certain key slides can all be recorded to make the job of the projectionist considerably easier.

This is easy to prepare on a recorder that has individual record switches for each channel. Many cassette recorders, though, have a single record switch for both channels. This involves some inconvenience in having to record the instructions and the sound track simultaneously. Of course, if you have an open reel deck (with separate record switches for each channel) available, you can get round this problem by recording on the open reel machine and copying onto cassette.

When using this method, on playback, only the channel with

sound accompaniment is connected to the amplifier. The instructions for projection are monitored by using headphones.

Automatic projection

Whenever a group of slide-tape enthusiasts gather together there is always the possibility of the outbreak of verbal warfare between the adherents of manual projection against those who prefer the automatic method. "Each showing is a unique individual performance" cries one side "hence the much greater chance of mistiming it", claims the other. "Your electronic devices are never completely reliable" retorts the first side, "but breakdowns are rare, and consider the saving in effort and anxiety during projection", replies the fan of the automatic method. Such arguments could go on for ever and, of course, there is no absolute answer. I am sure that both methods will continue to be popular.

There can be no doubt that a reliable automatic electronic fader can save a great deal of trouble and effort in projection. Timing, once perfected and recorded, will be consistent in all future showings. Automatic projection is, therefore, worthy of serious consideration. If more than two projectors are going to be involved, then automatic projection is, in most cases, essential.

The advantages of automatic projection should be self-evident. Of course, though, you still need a manual timing method to use while preparing the sequence.

Once again, there is no foolproof way of suggesting which method will suit you best. As the man said "You pays yer money and you takes yer choice".

Looking After Your Slides

When a slide-tape sequence is being put together it is very common to find a slide which, in content, is more or less what is wanted, but in other respects is lacking. For example, you may want a focal point (consisting of a highlight) to appear just to the right of the centre of the frame. The photograph available has the highlight dead centre or on the left or otherwise misplaced. A small change in position would have made all the difference when the highlight grows out of the preceding slide. It is, therefore, very common among the very keen enthusiasts to take three or four shots of the same subject (most professionals do this anyway, whether for slide-tape or not). This technique is often advocated and has much to commend it. It is cheaper to shoot an extra couple of frames rather than have to re-visit the original site to re-shoot, and possibly find that the conditions have become unsatisfactory.

With this in mind, you are likely to accumulate three or four times as much material as is needed for the sequence in question. This is not to say that three quarters of your transparencies will be thrown away, far from it. Save them all! They will, eventually, be usable in other sequences; even over-exposed transparencies can be utilised as sandwiches (photographic type!). But this topic is beyond the scope of this book. It is not uncommon for dedicated fans to collect sunsets, clouds, fires and trees etc., for use in likely future sequences. Some work on collecting material for several sequences simultaneously.

All this makes satisfactory storage and indexing of transparencies essential. If you already have a satisfactory system then carry on the good work; if not, the job can become a mammoth task if left too late, so read on!

Unmounted transparencies

This section obviously refers to those who process their own transparencies or who prefer their transparencies returned, unmounted, from processing houses.

Top quality transparency mounts are bulky (and expensive). It is much easier to file, store, and locate pictures on unmounted film strips. You can keep them in the sort of transparent envelope that is used in albums for storage of negatives. If the footage of film that you shoot in a year is small, then negative albums may suit your needs admirably. The very keen photographer may need something more suitable.

The problem with negative albums is their capacity. All right, before you protest, I know that most albums will hold 100 sheets and each sheet will hold (often) seven strips with six exposures on each. That is over four thousand transparencies, but for future cataloguing and convenience in locating a particular transparency it is not much fun wading through the pages of an album which could contain 42 completely unconnected images in each. Disconnecting and reconnecting the envelopes from the rings of the album is no joke either.

The solution to this problem is to acquire a filing cabinet (space permitting). Some good second hand ones can be obtained fairly cheaply from companies dealing with used office furniture. A home made drawer would suffice if other factors preclude a cabinet. The advantages of this system are numerous. Envelopes can be bought in bulk thereby saving considerably on cost, not only on the envelopes but also on the album; a single sheet can be reserved for, say, flowers and if you do a lot of this type of work, just red flowers or blue flowers. You may not need cataloguing, but if you do, it becomes much simpler. Each newly processed strip of film can be cut up and each image put in its appropriate place in your system; the location of a particular transparency simply involves identifying the appropriate (indexed) part of the file and extracting the relevant sheets for inspection. Another important advantage is that mounted transparencies can be treated similarly.

Mounted transparencies

A fair fraction of your transparencies will need a more critical inspection. Such problems as a slight camera shake will only show up on magnification. The best transparency of a few, each taken from a slightly different viewpoint, can often only be judged by comparing images on the screen. Hence mounting becomes essential for some shots. Once again, you can make a saving by using inexpensive card or plastic mounts. These are perfectly satisfactory for pre-judging purposes. Buying in bulk seems expensive at the outset but will offer years of supply, and at considerable savings (I have never known prices to go down!). The plastic ones have the advantage of being indefinitely re-usable.

Storing slides and isolating one particular example can be a nightmare for those without a system. Storage in the box in which they were returned by the processor is a common method but can be inconvenient. You can waste an enormous amount of time trying to find a particular slide and when you have finished with it – where do you stick it – sorry, replace it? You can, of course, reserve boxes for particular subjects – castles, people and so on; but stacks of slightly different boxes have a nasty habit of unstacking themselves. It is also highly likely that you will have far more subject headings than boxes.

The sort of slide box which holds one or two hundred slides vertically and separated from each other, and with an index on the lid is an improvement on the previous method but still offers problems. Individual slides must still be removed for checking, and replaced. Very time consuming. And remember, you may have a couple of dozen Tower of London or Statue of Liberty slides! The bulk and expense of twenty or thirty such boxes is no mean matter either.

I would suggest that, all other factors permitting, the filing cabinet technique takes a lot of beating. Transparent plastic files with about 24 pockets for 2'' × 2'' slides (and for other formats) are available. Some are designed to fit into hard-backed files (for storing on bookshelves) and others are fitted together with filing bars. These files can go into the same file as that for unmounted transparencies so that both slides and unmounted transparencies if, say, "trees" are kept within an inch or so of each other. Removing a file enables you to examine 24 slides or up to 42 unmounted shots at a time. The system, once organised, is virtually foolproof and, once established, has very few, if any, serious disadvantages.

The indexing of such a system needs some care otherwise you might find shots of the White House under both "buildings" and "Washington". Cross-referencing would be an advantage. If you have a special interest, say, in animals, then this section you can subdivide. Its your system, so its up to you to organise it in the best way. One point of advice, though, whatever you do, do not have a "miscellaneous" file. In these days of rush and bustle everything will be put in it — and bang goes your system!

Mounting slides for dual projection

Having chosen the slides for a sequence, performed a pre-liminary run, made amendments and finally approved the images, the time has come to mount the slides permanently. Obviously, the entire set of transparencies for each sequence (and preferably the whole programme) should be mounted identically. If you use mixed mounts then the image size on the screen may vary in area when changing from mount to mount (very small differences in area of aperture in slide mount can become significant in the greatly magnified image on the screen). Misregister, too, is likely to become a real problem. In addition, mounts vary in thickness which leads to some of the projected images being out of focus on the screen. Automatic focusing helps, of course, but is not ideal with electronic projector fading. If the auto-focus light fades with the main bulb for refocusing you have to wait until the image is projected onto the screen. This gives a momentary change in the image. Some people find this a little disturbing. It is easier to avoid this by using identical mounts.

What sort of mount should be used? Arguments bounce to and fro. The card or simple plastic mount offers the advantage of permitting free access of air to the transparency so that moisture equilibrium with the atmosphere is rapidly reached. However, such slides are subject to physical damage such as finger marks, scratches, etc., and, more annoying (at least on the screen) popping out of focus. From the standpoint of physical protection and freedom from popping, mounting between glass must win in spite of the potential of Newton's rings occurring (this can be minimised by using special glass with an "orange peel" textured surface known as "ANR" – anti Newton's ring) and, possibly less frequently, condensation occurring during projection.

However, there is one unquestionable argument in favour of mounting transparencies between glass and it concerns image quality on the screen. Most photographers are aware that it is nonsensical to spend a great deal of money on expensive, high quality camera lenses if the images so produced are to be projected with inferior optics (and this includes lenses for the enlarger as well as the slide projector!). However, what is not so well known is that projection of a non-glass sandwiched transparency with inferior projector lenses *appears* to give a satisfactory image over the entire area only because such an image is not *critically* sharp anywhere! The highest quality projector lens will emphasise any lack of flatness in a transparency so that sharpness will not be obtainable at (say) the centre of the image and at the corners simultaneously. It amazes me that people complain about their projector lenses when the real cause is a non-planar transparency in a cheap card or plastic mount. The same people, however, would recoil in horror at the suggestion that the pressure plate in the back of the camera should be removed! Optically, it amounts to the same thing.

The Kodak Ektanar-C projection lens is computed to allow for the average curvature in machine-mounted card slides. This is very effective with some slides; but unfortunately, even the most automated card mounting system leads to variation. So, this lens is no substitute for good glass mounts.

Probably among the best type of mount is the thin plastic, two-part mount with glass windows, which, when assembled make a sandwich of the transparency. Many of these contain a metal mask which is slotted to hold the transparency in place during mounting. The two halves are usually in two different colours, black or dark grey and white. It is important that these are used correctly. The white component is intended to face the projection lamp and reflect back as much heat as possible to enable the slide to keep cool. The dark half of the slide is intended to face the screen so as to minimise undesirable internal reflections.

Finally, such mounts do tend to be a little more costly than the average ones, but after all the time and effort you have put in to produce that little gem of a sequence, it would be a pity to spoil it for the sake of a few extra coppers, wouldn't it?

Mounting slides for precision in register

Very great precision in image register on the screen may be needed perhaps only once in every thousand fades, or less; but when imaginative fades are used successfully, the effect on the audience can be magnetic. The sort of images involved could be, say a number of colour posterisations of the same subject, the colours of which change as you watch the screen, or maybe a tree through the seasons; leaves (and blossom) developing in the spring, filling out in summer, changing colour in the autumn and becoming bare in winter. Such images not only require a special technique in camera work but they also need mounting with great precision, if the projected images are not to move during a fade and thereby spoil the effect somewhat.

Some precision mounts are manufactured which contain lugs which engage in the perforations at each side of the 35mm transparencies, but these will be of value only to those who use cameras which consistently produce images in exactly the same positions with respect to the sprocket holes in the film.

Those of us who cannot guarantee the image position on the strip of film, have to make personal adjustments. The problem with checking register with mounted slides is that of parallax. Two mounted transparencies placed one over the other may appear to be in perfect register, but, due to the separation between them, will seem out of register if the viewing position is changed somewhat. (The same effect can be observed by closing one eye and lining-up a near and a distant object, such as a vertical window frame and a similar one across the road. Opening the other eye and closing the first one, without moving the head will cause the two objects to appear out of alignment).

Two methods of overcoming this problem present themselves. The first involves eliminating the separation between the transparencies. The simplest way to achieve this is to mount one of the transparencies (we will call this the "master" and should contain well defined details which can be used for checking register) normally in the black half of the mount, ie. that which will face the screen. Even if this mount contains slots in the mask to hold the transparency in position, it is worthwhile ensuring lack of movement by anchoring two opposite corners with thin adhesive tape. The second transparency is now placed the *wrong way round* in the white half of the mount so that when the two parts are placed together the transparencies will be in contact and the images will almost coincide.

Adjustment, to the non-master, by trial and error, enables satisfactory register to be obtained. This is worth checking with a magnifier before the white part of the mount is pressed in place on the second mount. Any other transparencies needing to be registered can be compared with the master in the same way before the latter is mounted permanently. It is not advisable to register slide 3 with slide 2, then slide 4 with slide 3 etc. as any errors in register may be cumulative.

The second method is worthwhile adopting only if you expect to do a great deal of close register work. This involves using

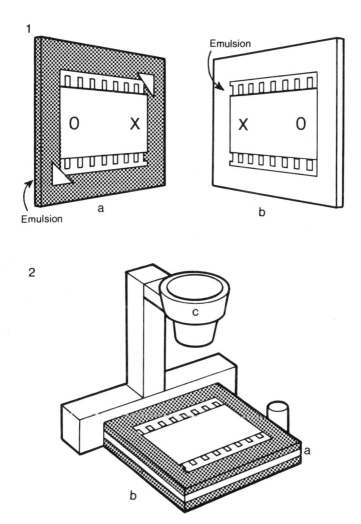

1, a, Black half of 'master', transparency fixed; b, white half of second (or third etc.) slide, transparency movable for adjustment. 2, a, 'Master' transparency; b, light table; c, magnifier and eye-cup mounted exactly over centre of transparency.

one transparency as a master, as before, but this time its mounting can be completed. The transparency to be registered is placed in the black half of the mount and is then placed over the master. Both are placed on a light table (see page 103) and are pressed against a suitable stop at the top edge and the right hand edge of the mount (these are the edges which rest against the stops in the projector when the inverted slide is in place). The slide can be viewed during its positioning. Parallax is avoided by viewing through a magnifier clamped in such a position that its optical axis passes perpendicularly through the centres of the two slides. A rubber eye-cup, such as those available to clamp on the viewing window of single lens reflexes, fixed above the lens can ensure that the position of the eye does not vary. Lens alignment can be checked by using two, accurately mounted copies of Appendix 2 placed one above the other, but separated by two or three empty slide mounts. The position of the lens should be moved until the fine lines crossing at the centre are superimposed perfectly and the lens can then be fixed permanently into this position.

Spotting and numbering

Have you ever been to a slide show where the occasional slide is projected upside down? Audiences are usually kind if only one mistake like this is made but they can be raised to the heights of mirth if it is repeated often. Vertical sunsets usually bring the house down especially if a weak voice from the projectionist calls "sorry about that folks".

No projectionist wants to be thus reduced to a sobbing mass and no photographer enjoys having his show highlighted by repeated gales of hysterical laughter. Nor is there any reason why this should happen. Even if you drop a slide tray just before a show and scatter slides about the floor, the show should still go on, and the audience should be unaware of any mishap.

The secret lies in careful spotting and numbering of slides.

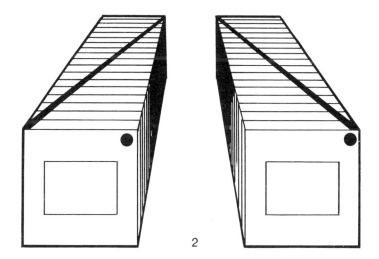

1, Spotting slides. When held the 'right way' the spot should be placed in the bottom left corner. 2, Diagonal lines on slides forming an 'arrow' pointing to the screen assist positioning of left and right magazines.

First of all, spotting. It is as well to acquire specially made spots of two different colours, one colour for each projector, say red spots for the right hand projector and yellow for the left. The positioning of such spots is also important, there are eight corner positions available, seven of them being unsatisfactory. If the slide is held so that the image is the right way up and the right way round then the spot should go in the *bottom left* corner. When in position in the slide magazine the spot will be visible in the top right hand corner when looking towards the screen. A quick glance can detect whether a slide is in the correct magazine or not.

Another useful exercise is to paint a diagonal line across the upper edges of the slides in each magazine. This assists the recognition of any slide which may be positioned incorrectly. If the line travels from rear left to front right with magazines intended for the left projector, and vice versa, then matching of magazine and projector is facilitated. (Nitrocellulose lacquer used for automobiles or even nail varnish has the advantage of drying quickly, but do not use too much – for your slide's sake!)

The numbering is best done when the slide is inverted so that each number runs in sequence in each magazine. Some workers prefer to have odd numbered slides in one magazine and even numbered ones in the other. Others prefer consecutive numbers in each magazine. It is a matter of personal choice. (However, the organisers of some competitions insist that odd numbered yellow spotted slides go in one magazine and even numbered red spotted slides in the other. A blank slide should be numbered zero.) If manual fading is involved, then a blank slide in the first magazine is sufficient. When automatic projection is to be used, then sometimes there can be an advantage in putting a blank slide in each magazine, especially if the "pause" button on the recorder is to be used to start the programme. The reason is that, with some fading devices, both projectors operate at half-brightness when the recorder is on "pause", so that the first slide of a sequence is seen on the screen before the sequence starts properly. Naturally, the last slide of a sequence should also be a blank to provide a fade-out, and this also should be spotted and numbered properly.

Storage conditions

No dye is completely stable to light or resistant to colour change under adverse conditions and those used in colour transparencies are no exception. It is, therefore, advisable to provide as near to the ideal storage conditions as is possible. Fortunately the conditions required are not too difficult to achieve. Dampness is the arch enemy of the colour transparency so that relative humidities ought not to be allowed to become too high, about 50%–60% is satisfactory. Temperature is also important and the fading of a dye, like any chemical reaction, is accelerated as the temperature rises. A temperature of 20°C (68°F) should have no adverse effects on slides. It is worthy of note that high temperatures coupled with high relative humidities can not only cause fading but also, various microorganisms such as bacteria, moulds and fungi find the gelatin of film ideal nourishment! Any adverse effect which they produce is irreversible. Other sources of potential trouble include paint fumes, wood stains (and even freshly cut wood) and chemical fumes in general. "But", you claim, "you don't get chemical fumes anywhere near slides, in the average house". Don't you believe it! What about those transparent plastic files? I know that plastics are more or less inert, but they are often fairly brittle so that plasticisers are incorporated to make them more flexible. Here is where the trouble lies. Over a period of time, plasticisers can migrate out of the plastic and, although their volatility is very low, traces do evaporate – (onto your slides?) Even though I am unaware of any indisputable evidence of slide deterioration due to plasticiser vapour it might be thought safer to choose the more rigid files rather than extremely flexible ones.
If the storage conditions are likely to vary a great deal then the only sure way to protect your slides is to seal them in metal boxes, and, as an extra precaution include some moisture absorbing substance such as silica gel.

Putting a Simple Sequence Together

What is the object of making a slide-tape sequence? There could be a dozen answers to this question, all different. One thing, however, is common to all, and that is to entertain, in one way or another. Is there any point in spending lots of time, money and effort in producing a sequence and then hiding it away from the eyes of others? Possibly, if the photographer derives enough pleasure from its creation. However, we all thrive on the praise of others and, for that matter, their criticism, for only with the latter can we hope to constantly improve the quality of our work. Hence entertainment must rank high, if not highest, in the list of objectives.

When we consider the variety of interests and attitudes of people in general we may ask how is it possible to please all of them all of the time. The answer is, of course, it is impossible. You can though offer each person something of interest provided you include a variety of material in a performance. Let us assume, then, that the long-term aim is going to be to produce entertaining programmes containing sequences of different types and content. All right, but where do you start? For your first sequence you need something which is not too taxing, something which will whet your appetite, something which will enable you to get a feeling for the medium.

Now, which comes first, the images or the sounds? Perhaps neither, maybe it is the idea, a story, a message. You must decide. There are many who claim that we are, primarily, photographers and hence the images must be the more important. I claim that this is not always so. Admittedly, the images must be as good as you can produce, but they need not be an end in themselves. I can recall sequences consisting of photographs which had no outstanding merit, certainly not individually, but with the appropriate sounds produced an impact which caused them to be remembered years later, long after others containing superb photographs had been forgotten. Similarly, we cannot be too dogmatic about the sound. It is not an end in itself. It is a part of the whole. Surely, the essence of a good sequence is the idea, the blending of images and sound to produce the desired effect.

I am going to assume that mixing of sounds, and that includes the human voice, is not to be involved. The first stage is get some experience in arranging slides in a suitable order, finding the right sounds and blending the two together to produce an entertaining sequence. Many of the more advanced workers in this medium tend to look down on such sequences. Pretty flowers accompanied by pretty sounds, they claim, do not do justice to the medium. But does this matter?

I have yet to meet a person who actually dislikes such sequences, providing they are well produced and are not too long. (This last point is most important and more will be said about it later.) No, choose a subject which will give *you* pleasure. Pay no heed to the critics on this score. They may be right, but you will discover the limitations of simple sequences in your own good time. It is the old story of walking before running. If you fall at the first hurdle, you may never want to continue again!

To describe the process, I need a suitable example. Here are two possibilities, a sequence on flowers and one on London. (There are hundreds of other subjects, choose whichever you like, but for obvious reasons, I can make reference to a few only.) Perhaps you have a good collection of slides on your subject already. Good. But you are likely to need more, so choose a subject for which you can shoot more material as needed so as to fill in the gaps.

Now to find the sounds. The flower sequence needs suitable pastoral music; but for goodness sake avoid Beethoven's Pastoral. There is nothing wrong with it, naturally, but it has been done to death, somewhat. Lesson number one, try not to be too hackneyed or predictable. Rummage around your local record library and find something less well known.

People will come along at the end of the show and ask what the piece of music was. It's all a part of the technique of entertainment. (Few of us are over-impressed with the comedian who tells stories we have all heard before.)

The London sequence needs something with an obvious flavour, maybe with lyrics, and these will make further demands on the sequence. For example, if the lyrics refer to cockneys, there is not much point in showing photographs of boats on the Thames! The images should link with the sounds.

When choosing your sounds note that they impose their own restrictions. The first obvious one is that of duration. Do not be misled into thinking that fifteen minutes of enjoyable music and the same period of photographs of beautiful flowers will produce a quarter of an hour of satisfying sequence. It most probably will not. (Unless you do it with a great deal of skill.)

Better to be on the safe side and aim for about three minutes, at least for a start. Now you have to find not only a suitable piece of music, but also one of the appropriate duration. Here is where the difficulty lies, the choice of music is very much more limited. One answer is to choose a passage which almost fades away before the next section is introduced. Such music could be faded out completely at this point in order to terminate the sequence.

Songs of London of suitable duration are likely to be easier to find. In fact, two short songs could be linked together if they relate to different facets of the subject. In this way the duration could be extended to, say, six minutes without necessarily introducing an element of boredom.

At this point, you have sounds and a collection of slides; you are ready to put the sequence together. A very quick way to become frustrated is to flash the slides onto a screen, one by one. Trying at the same time to jot down which ones appear to suit the purpose. No, the most convenient way to select slides is to use a light table.

The light table

A light table is simply a flat surface illuminated from the rear on which a number of slides can rest. At its simplest, it can consist of a sheet of glass with a diffuser of tissue paper attached to the rear side, resting on two chairs and with a table lamp beneath it. More permanent light tables often have the window/diffuser tilted at an angle and have horizontal ledges on which the slides rest.

There is little difficulty in making a light table. A sheet of translucent plastic can act as the diffuser. If you want a sheet larger (in either direction) than about two feet, then you may need a backing of plate glass to provide sufficient rigidity. The ledges on which the slides rest can be made from $\frac{1}{8}$ inch square cross-section wooden strip available from model makers shops. Spacing should be about $1\frac{3}{4}$ inches apart (if greater than 2 inches then the slides drop in between the ledges and become more difficult to remove). A PVA latex-type of adhesive is usually satisfactory for fixing the ledges to the plastic. Illumination by a fluorescent tube is convenient and minimises temperature rises. The light table should be as large as space permits, but if it is of the horizontal type a dimension from front to back of more than 30 inches makes close study of the most distant slides rather difficult as well as being a little inconvenient to reach. Personally, I prefer a large, horizontal table on which slides can be spread out and an adjacent tilted table for the selected slides to be placed in order of projection.

Editing the slides

There is one inviolable law for editing slides. Be ruthless! There is no excuse for showing slides which are in any way inferior. Slides which suffer from technical imperfections, such as those which are out of focus, over and under exposed or suffering from camera shake must be rejected mercilessly. There is rarely any excuse for faulty images (possibly if you are shooting a sequence on mountaineering while suspended at the end of a rope over a 1000 foot drop, with 60 mph gusts of wind, a certain amount of camera shake may not be unduly criticised — but still try to keep them sharp, nevertheless!)

At the light table, do not reject a slide prematurely because of poor composition, it may fit neatly into a series of fading images. The requirements for cross-fading pictures can often be quite different than those intended for individual viewing. Indeed, it may be advantageous to have a focal point of a limited number of slides alternately at one side of the screen and then at the other in order to aleviate any element of monotony.

The process of arranging slides in order, for projection, is much easier to demonstrate than describe. There are so many factors to bear in mind. For example, if the musical accompaniment flows fairly evenly then the choice of order of slides is fairly flexible. If, on the other hand, there are dramatic peaks in the music, crashing chords and so on, then the overall impact will be increased if suitable dramatic images accompany them. Hence there will be certain key points in the sequence where the requirements of the images will be dictated by the sounds. Even the colours and nature of the photographs can be governed, to some extent, by the nature of the music. Soft, flowing, pastoral passages are ideally matched with peaceful colours, soft greens, muted lavenders and blues. (The rates of slide change and fade should match also.) Busy sounds are more suited to photographs with much detail and more active reds and oranges — and more rapid fades and slide changes. An exercise which is worth trying is to project a mixed collection of slides and decide in each case whether a sound accompaniment would best be "busy" or gently flowing. Discuss your ideas with your spouse or a (sympathetic) photographic friend who has a feeling for aesthetics. You might be surprised at the outcome! The overall aim is to match images and sounds to form a coherent effect.

Assuming, then, that any key points in the sequence have been looked after, what criteria do we use to arrange the

Above and opposite: Examples of images which match 'soft' and gentle music. Cross-fading can be slow and leisurely. (*Above,* photograph by Sue Duncalf.)

Plate 1: Violent or active music can accompany images of this type. Rapid cross-fades and snap changes can be consistent with the mood evoked.

Plates 4-5: Considerable impact can be obtained at the opening of a sequence if a posterisation fades, in register, into a 'straight' photo.

Plate 6: Animation effects can be achieved with two projectors provided that the effect is kept fairly simple and you do not attempt to compete with cine.

Plate 7: A highlight, which often commands attention of the viewer, can be in one area of the frame and the remaining subject material moved around it.

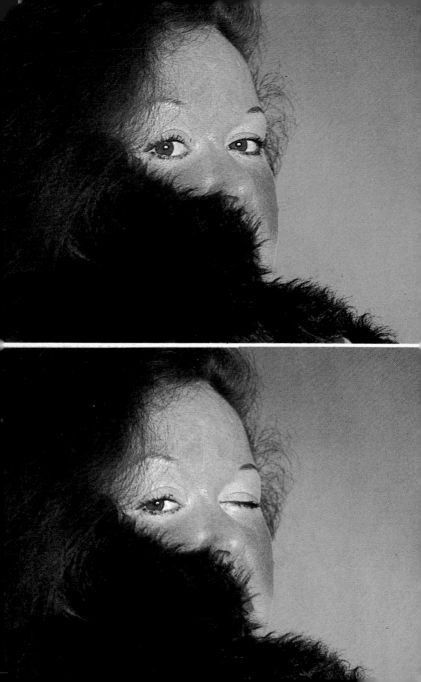

Above: To the audience, the wall with the 'peephole' appears to dissolve away before their eyes. The photography for this needs considerable care as the central unchanging subject must not only be exactly the same size in each photograph, but must also be in exactly the same position in the frame. (Eric Cooke)

Opposite: This three slide change requires the camera to be tripod mounted if register on the screen is to be maintained, otherwise the whole effect is lost. It is quite possible for there to be only two occasions in the whole year when the sun is in the precise position to produce highlights like the one shown. (Eric Cooke)

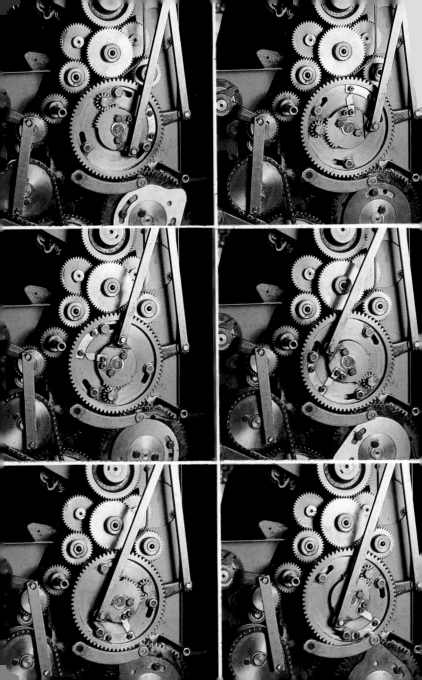

Above: Humour, often difficult to catch in a single photograph, can be injected into a series of photographs, especially with suitable accompaniment. Here, if the change is made rapidly, the audience will be unaware of the change in background. The sound of falling rocks and crashes completes the illusion.

Opposite: The study of the behaviour of cam, levers and other moving parts of machines can be made effectively by the use of two projectors. Rapid cross-fading of several slides can illustrate a full cycle of operations.

Plate 12: The change from long shot to close-up can be most effective when using two projectors, especially if the centre of interest is kept in the same position.

Plate 13: The attention of the audience can be led around the screen by bringing a highlight out of a shadow area. With this fading pair, the attention is brought to the bottom left hand area, at least temporarily.

Plate 16: Ending with a sunset can be a sure winner, as long as it is not done too often. Extra impact can be obtained if the images are kept in register on the screen.

Plates 14-15: Changing the point of focus is an effective technique, but should be used with some restraint. It renders the out-of-focus object virtually unidentifiable, but cross-fading from one to the other says 'keep out' as effectively as any noticeboard.

remaining slides? A number of avenues are available:

Brightness. It is unwise to indiscriminately alternate dark images with bright ones. The eyes of the audience have to constantly re-adjust. Better to keep bright high key images together, and dark, low key images in another part of the sequence, unless you have a definite reason for not doing so.

Colour. Colour provides an obvious way to link separate images together. Changes of, say, from red to blue to yellow to violet give the impression that the slides have been put together without any serious thought. Changing colours through the spectrum, red to orange to yellow etc. are often more pleasing to the eye and also show that there has been some design behind the order in which the slides have been arranged.

Seasonal changes. Unless there is some overpowering reason for not doing so, showing the seasons out of order will claim little merit. A keen horticulturist in your audience will, very quickly, be made aware of your lack of attention to detail if you put roses before daffodils and fruit before blossom.

Shape. One of the advantages of cross-fading technique is that an object of a given shape can be made to dissolve away and be replaced by a different object but with a shape similar to the one that preceded it. Thus, the sun can become a sunflower, a single autumn leaf against a blue sky can fade into a branch of such leaves and eventually the full tree. Like all effects, it can be overdone and you must constantly be aware of excessive use of technical ploys. The use of shape, however, can be a most useful tool to link one section of a sequence with another having a different emphasis.

Texture. Although the use of texture as a linking theme is less common than previous examples it offers yet another means of selection and presentation of a series of slides. Depending upon what you intend, you can compare or contrast textures either with a series of slides or within each slide.

Movement. If the sequence deals with people or animals, the order of showing can be arranged to follow the progression through some activity. A documentary on a festival or procession can show, for example, the brass band approaching the camera position, on a level with it and then leaving it behind (if facilities are available, stereophonic sound can be made to complete the illusion of movement by synchronously panning the sound with the images).

The suggestions above are intended to offer some ideas for arranging slides in some sort of order, but at the risk of becoming vague and imprecise, such systematic arrangements must not be overdone. As with all things, excess is as bad as abstinence and the over-use of any of the ideas surveyed above will result in monotony. Like the "rules of composition" they are merely suggestions to help.

At all times, be aware that following a specific technique slavishly simply leads to predictability and results in monotony. Your aim should be to make each member of your audience maintain his attention on the screen lest he misses something enjoyable!

Horizontal and vertical formats

One of the most common mistakes made by beginners in this medium is mixing formats indiscriminately. Generally speaking, the cross-fading of horizontal and vertical slides tends to produce a cross-shaped image part way through the fade and the effect is usually rather displeasing. This is not to say that such changes in format cannot be achieved satisfactorily; they can, but to be successful, certain techniques usually need to be adopted. These will be discussed later (see page 134).

Most photographs are taken in the horizontal format and many slide-tape enthusiasts tend to shoot with this end in mind. However, some subjects,such as tall buildings, are best suited to a vertical composition, so that if a sequence is likely to contain many such photographs then the best measure is to collect a series of vertical shots and make a separate section of them within the sequence.

Origination of title slides

Title slides add a touch of polish to a programme of sequences and your audience has a right to know what to expect. There is a world of difference between showing a single sequence to a few friends at home (where you can explain what you have in mind) and showing a programme to a large number of people, many of whom may be strangers. Title slides (and slides indicating the end, if you wish) provide a means of isolating one sequence from another, or, if required, one section of a sequence from another.

While good title slides can add class to a programme, poor ones can have a considerably adverse effect. Consequently, unless your graphic skills are well developed, it would be best to use a professionally produced text matter. This is not to say that you need employ a graphic artist, other methods are available. The eventual products are photographs, but you are likely to do the preliminary work on paper or card.

Typescript

Most people have access to a typewriter so that little difficulty is likely to be encountered in producing title slides from the typed sheet. Unfortunately, typescript is rather limited in suitability for many purposes. It could be ideal for a documentary on the life of a girl in the office, but quite unsuitable for a story about the same girl and her fiance.

Transfer lettering

Nowadays there is an ample supply and variety of transfer lettering in shops which cater for artists and graphic designers. It can be almost guaranteed that you can find a suitable typeface for virtually any form of subject.

The method of application is simple. Place the part of the sheet containing the letter to be transferred over the appropriate part of the paper, or better still, thin card. While holding the sheet firmly in place, rub a ball point pen firmly over the wanted letter. That transfers the thick ink (of which the letter is composed) from the reverse side of the sheet or substrate. Gently peel away the top sheet containing the remaining letters, and burnish the letter into place with a suitable instrument such as the convex part of the handle of a teaspoon.

You carry on in the same way. One letter at a time. The main problem is in keeping the lettering exactly level and evenly spaced. The broken horizontal lines on the lettering sheets are intended to help here. Line them up with the edge of your card, or with fine pencil lines. The length of the dashes represent the recommended spaces for each letter immediately above.

Transfer lettering is, in most cases, manufactured in both black and white, so that choice of the correct colour can eliminate the need to reverse tones photographically.

Purpose designed lettering. Occasionally the commercially available typefaces may not be quite what you want. In this case, an existing typeface will have to be modified or a special face designed for the purpose. Rough out some initial ideas in pencil and then attempt to perfect the most promising one. If it seems to be beyond your skills, then try someone with some artistic ability. Your rough drafts will help to illustrate what you want.

Producing title slides

Although it may be one of the first slides to be shown, the title slide could well be the last to be produced. There are some advantages to organising affairs in this way. For example, perhaps the title is intended to be projected, temporarily, over an existing image on the screen. This means that the title must be in pale colours and must be projected over a dark area of the existing image. (Black images cannot be projected satisfactorily over existing pale images). This being the case, the position of the text on the title slide may be critical. If dark text matter is required over a pale background then the text is best arranged on the actual slide of the background. In order to obtain sufficient contrast for title slides, use line or

Remove backing sheet and with Letraset on work surface rub down first letter with old ball-point pen, remove Letraset sheet. The first letter has now off-set on to work surface. Lay backing sheet over letter and gently burnish with bone handle. Position Letraset sheet so that second letter can be impressed. Second letter is now off-set on to work surface.

lith film. Both of these produce very dense blacks and clear film for highlights. For black lettering on a pale background you can sandwich a high contrast positive with the appropriate slide. You can either project the sandwich itself or copy it onto colour reversal film.

Two methods of producing the text on film are immediately obvious. One is to prepare a negative image on card (white lettering on a black background) and photograph this using the high contrast film in the camera (the "speed" of lith film is in the order of 1–5 ASA). Development gives the positive image directly. The second is to produce a positive image on paper or card, photograph this onto continuous tone film (there is advantage in underexposing and overdeveloping to increase contrast) and contact print this onto high contrast film. Whichever method you choose do not approve the end result until you have checked the *projected* image (high contrast films are notorious for producing pinholes in dark areas and these will need to be retouched).

If you want a pale text on a dark background, two techniques offer themselves. One is to use one projector for an image which remains on the screen, and the second to superimpose the text matter temporarily over a dark area of the existing image. The text may be white or coloured. If coloured, a number of secondary colour effects can be produced, white, for example, can be obtained by superimposing red and cyan, green and magenta, or blue and yellow. The table on page 166 indicates some of the colour effects possible.

The second technique for pale text on a dark background is to combine both on a single slide. Again, several methods are available. You can arrange the pale lettering on glass or any other suitable transparent material and then photograph it in front of the background material. You may need a small stop or a split close-up lens to take care of the depth of field problems. Where the background material is very small or otherwise precludes this technique; you can project two slides (one of the background and the other of the text) onto a screen and the combined images photographed in colour. Of course, you need a suitable film or colour correcting filters.

Recording sounds

First, let us see how not to approach the problem! Whatever you do, do not play your music over speakers and record the sounds so produced via a microphone onto tape. This method is fraught with difficulties. Firstly, the quality is bound to be inferior. The sounds you record in this way depend among other things upon the acoustics of the room in which the recording is made, the quality of the speakers and of your microphone. Furthermore, the sound of passing traffic, a knock on the door, the budgie chattering (as well as your language directed at the perpetrators of such crimes!) is all recorded. It only takes one such experiment in recording to show us how much our ears and brains subconsciously reject sounds which are present, but which are of no direct interest to us.

The best way to record from a disc or tape, is directly onto tape. The leads from the record player or the tape recorder can be connected directly to the recorder. (If recording from tape onto tape, obviously you need two tape machines.) This method excludes any external sounds. Occasionally the signals obtained from a record player are not strong enough to record at a satisfactorily high level on tape. If this is the case then the signal from the record player is best fed into the amplifier and the output signal from the latter fed into the recorder through the appropriate terminals. The normal 5 pin (DIN) connection between tape and amplifier carries both record and playback leads.

When recording, ensure that the maximum signal is recorded, consistent with sound quality. If the signal recorded onto tape is too weak, then, on playback tape hiss will probably be objectionable. Aim to record in such a way that, in the loudest parts of the music, the VU needles only just enter the red regions of the scale. If they deflect well into this part, then distortion may be obvious on playback. Learn how your particular recorder reacts to different recording levels before you make an important recording.

The best way to get quite accurately timed recordings is to

use the tape recorder pause key. Select "record", start the tape moving, and stop it at once with the pause. Now start the record player. Wait until the section of the record you want is about to start, then release the key. That way you should avoid clicks from either the recorder or the record player. At the end, either fade out the music (with the record level controls) or stop the tape again with the pause key.

Editing sound

Suppose you have now made a suitable quality recording. Perhaps two pieces of music, one after the other. You have faded one out after the appropriate length of time and have built up the volume of the second from zero to the maximum required. The latter may end naturally, or you may have faded it away in order to end the sequence. Perhaps, though, the interval between the two is too long and maybe, there is a recorded clunk as you switched the record player off before changing from one disc to the other. These faults will need to be rectified before the taped sound is in its final form. This is done by cutting out the unwanted region of the tape and splicing the remaining ends together.

Locating the positions for cutting is the first task. If you can play the recorder with its protective head covers removed, do so. It makes tape marking and editing much easier. Play the recording through and make a mental note of the point at which the wanted material ends. Note the footage indicator either simultaneously or after a re-run. At this point it will become obvious that the tape speed for recording has a considerable effect on the ease with which editing can be carried out. At a recording speed of $1\frac{7}{8}$ ips (inches per second) a $\frac{1}{2}$ second playing time corresponds to a length of tape less than one inch, so that the position of the first cut will have to be very accurate. At $7\frac{1}{2}$ ips the same duration of playing occupies over $3\frac{1}{2}$ inches; so that small errors in cutting are much less serious. Such matters become less troublesome as you gain experience in editing. Even so, the higher tape speed is to be preferred as this will give a better sound quality (professional

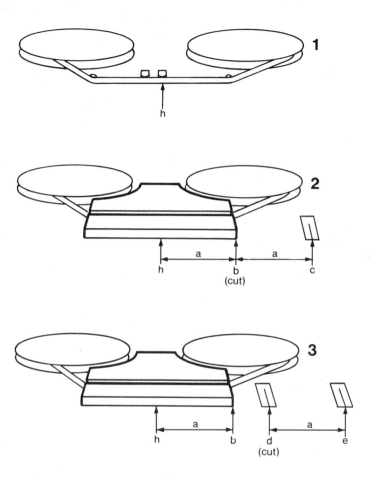

1, Head covers removed to assist location of playback head (h). 2, The distance (a) between the playback head (h) and the end (b) of the head cover is noted. The label (c) is placed the same distance away from (b) as (b) is from the head. When the position of the cut is located on playback, the tape is stopped, gripped at (b) and pulled to (c). The cut can be made at (b).

3, Alternatively, the labels (d) and (e) are placed the distance (a) apart. After locating the position of the cut (h), the tape is pulled until the grip at (b) reaches (e). The cut is then made at (d).

studios normally record at 15 ips). Once again, rewind the tape, then play it back and use the pause control to stop the tape at the precise position required for the cut. Turn the tape movement control to off – or switch off the power. If it is still in place, lift up the head cover. Mark the position of the *playback* head on the tape with a wax pencil.

Now move on to the beginning of the next piece and mark the tape again. Once the position of *both* cuts are marked, pull the tape clear of the heads and you are ready to cut it.

There are some variations on the method described above. For example, you can fix an adhesive label or any other suitable mark on the right hand side of the tape deck. Place it so that the distance between the mark and the right hand side of the cover protecting the heads is exactly the same as between the same point on the cover and the playback head. When you have stopped the tape at exactly the right point, grip it right next to the right hand side of the head cover. Then pull it out until the gripped part of the tape is over the mark or label. The position of the cut is then indicated by the end of the head cover. This method avoids the use of wax pencils and the need to remove the cover from over the heads.

The accurate location of the position of the cut is sometimes difficult with the normal tape transport mechanism. A more accurate way is to stop the tape with the pause control, (this simply detaches the capstan drive from the tape but maintains the tape contact with the heads), and move it by hand. Using headphones for monitoring, the tape can be moved forwards or backwards over the heads by rotating the tape spools. The pitch of the sounds through the headphones is altered, naturally, but with practice, you can find the required point of the tape very accurately. (For a practice run for those trying this technique for the first time, recording sounds at the slowest speed offers the advantage of ease of aural location of the position for cutting.)

The second cut in the tape can be made with little further trouble provided that the amount of tape to be rejected is only a foot or two. If, for any reason, there is a considerable amount of tape between the first and second cut, and

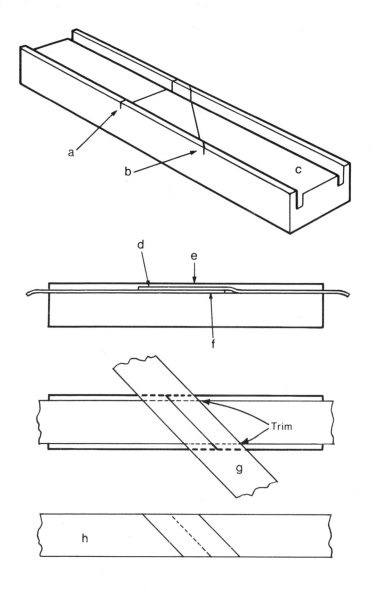

A splicing jig is used. a, Slot for 90° cut; b, for 45° cut; c, groove for tape; d, remove after cutting; e, cut; f, discard after splicing; g, adhesive tape. h, The finished splice.

especially if you want the recorded material on the middle length, then this is best wound on to a separate spool (and the spool labelled) before the second cut is made.

At this point you have two lengths of tape, one on each spool, and the ends need joining. It will be false economy to stick them together with any adhesive tape which you first lay your hands on. There are special splicing tapes available for the purpose. They do not leak their tacky adhesive and thereby avoid soiling the tape heads, pressure pads and other parts of your valuable recorder. Furthermore, special splicing jigs are sold, which make the task of splicing very much more simple than the method of using scissors. They do not cost much and are a worthwhile investment.

The method is simple, lay one end of the tape, coating side down, in the channel of the jig. Fold down the pressure pad to hold it in place, lay second end over first, held down with the other pad. Align the two wax pencil marks accurately. Now cut through both tapes simultaneously at an angle of 45° with an unmagnetised razor blade. Follow the blade guide in the jig. This provides a butt-joint. Lay the splicing tape over the joint, parallel to the cut, press it into position. Trim the ends (using the guides in the jig) to finish level with the edges of the magnetic tape.

It is apparent that some excess of magnetic tape is needed for the purpose of splicing so that the initial cuts in the tape are best made, say $\frac{1}{2}$ inch after the first section and the same distance before the second section. The only other equipment I would suggest is a plentiful supply of razor blades. After about six or so cuts, the blade loses its edge (at least at the corner!) and nothing is more annoying than trying to make a precision splice with a blade which tends to pull and tear the tape. (When the corners have been rendered useless for cutting magnetic tape, the blades are still perfectly satisfactory for shaving!)

As a check on the quality of the splice, the tape can be rewound and played back. If the job has been done properly, then the ear should be unable to detect where the splice occurs. If it can, then go back and do it again!

It is often useful to mark the start and finish of two or more special recordings on a tape. To do this you splice in short pieces of coloured leader tape. If you want to be accurate, you can standardise the distance between the end of the leader and the exact beginning of the recording, for example, to the distance between the replay head and the final tape guide post on your recorder.

When you are working on tape-splicing or recording – small pieces of paper make excellent temporary markers. Just pop a fragment between the coils of the take-up spool as the tape winds on to it. You can go back to the paper easily. It saves writing down counter numbers too, if you want to mark several places.

Copyright

Before leaving the section on taping, the subject of copyright must be mentioned. Any person who copies from disc, tape, film or (usually) radio etc. is, strictly speaking, infringing copyright and can have legal action taken against him. Everybody is aware that such copying takes place extensively for private use, but this still does not in any way detract from the legal implications. While it is unlikely that action will be taken against an individual who copies material for his own private enjoyment, once such copied material is used in public, then the copier leaves himself wide open to legal repercussions.

Such defences as ignorance or the lack of financial gain on the part of the copier are quite valueless. The only way to avoid complication of this nature is to obtain permission in the form of a recording licence which can usually be obtained for a small fee. This subject is dealt with more in Appendix 7.

Timing the slides

This is where slide-tape gets interesting. You stand before the light table covered in slides (the light table, that is!). A stop-clock is nearby, the recorded sounds are ready to be played back and you have a rough idea of the rates at which slides are going to be changed. Switch on the sound and start the stop-clock at the appropriate point (eg. the title slide, if this hasn't been made yet, then insert a blank for the time being).

Listen to the sound while looking at the slides and every seven or eight seconds move to the next slide. At the end of this exercise it will be apparent that the timing needs altering. Perhaps that crashing chord came too early for that key slide with impact. The rates of slide change preceding it will need to be speeded up accordingly. The times for each slide change can be marked on a cuesheet, or even in pencil on the slide mount.

Go through the sequence again. Yes, its getting better. What about those ten slides in the second minute? They are timed to show at exactly six second intervals. That is going to be monotonous. Those times and rates of fade need revision. Some can have eight seconds on the screen and others four seconds. Which ones are going to have four seconds? Not that one, it's too good for such a short showing. Nor that one. Wait a minute, they are all superb slides, they all deserve more time, not less. Too bad, some will have to go. Again, be ruthless, not because of lack of quality this time, but simply because there is not room for them all in this sequence. This time reject the slides which do not appear to give such a good fade. It could be worthwhile projecting them to check whether the cross-fades live up to the effect anticipated on the light-table.

The timing is finally right. You have marked the mounts with the times and prepared a master cuesheet with all the times of slide changes. Now you can load the magazines and see what it looks like on the screen.

The initial showing and amendments

This is going to be the real test. But how are you going to watch the screen and the cuesheet and the stopclock. It is not possible. Get somebody to help you then. What, at two o'clock in the morning? All right, give it a rest and finish it tomorrow. Not likely, not when I'm so near. So get your slide-tape synchroniser and put some blips on the tape at the appropriate times. It will only take as long as the duration of the sequence.

After all that effort, it looks pretty good. Not perfect, but certainly not bad for a first try. The best thing to do at this stage is to take a break and come back to it fresh later, perhaps tomorrow. But I would not mind betting that you wake up the next day with it on your mind, and with some ideas for improvement.

Well, at long last it's finished. A further one or two minor changes needed perhaps, but its progressed as far as you think it can go. The problem is, you have seen those slides and fades, and heard the sounds so many times that you are virtually sick of the whole thing. By now, you can not decide whether its good, bad or indifferent. Somebody else will have to help in this. By that I do not mean that aging, doting aunt who thinks that everything we do is "lovely" and "beautiful". I mean someone who is sufficiently critical. A fellow photographer, especially one with an interest in this medium, is the sort of critic you want. Not that you need to take all of his recommendations, he may have missed the point you are trying to make. Nevertheless, if he is a reasonable person, some of his comments will be valid and you can act accordingly.

Before we finish completely with your simple sequence, let me make a prediction. After your sequence has rested on the shelf for a few weeks, when you see it again, its limitations will be obvious to you and during that rest period more ideas for more sequences will have captured your imagination. Maybe a documentary or a gentle jibe at one of your pet hates has occurred to you, or perhaps a record you heard recently suggests something to which you can give a subtle twist. Of course, a single piece of background music is going to be utterly inadequate, you are going to need something with more scope. Photographically, it's going to be taxing too. You need to put that particular point over very carefully. Ah well, here you go again! Once you get bitten by this audio-visual bug there's no escape. You may as well start ordering your colour film in bulk.

Getting More Ambitious

It's a good feeling to see your friends enjoy your sequence, or better still, sequences. (By the time you have perfected a sequence you will probably be sick of the sight and sound of it yourself!) Listen to their comments and go further. Solicit their *honest* criticism and pay heed to it. By this, I do not mean that you should give undue attention to the character who says "I don't like flowers, why don't you have more sports shots or girlie pictures?"

He has a point, of course, but this is concerned with variety in a programme rather than constructive comments on any single sequence. There will always be those in your audience who do not like flowers, or abstracts, or landscapes, or animals or a thousand other things. But it is the quality of each sequence which matters. Does it say what you wanted it to say, or achieve any other aim which you had in mind? Only your audience can answer such questions. If you get the same criticism independently from two or more of your audience then make a careful note of their comments, they will be valid. You may not agree with them, but the comments are valid for all that. (I am assuming here that your aim is to improve the quality of your work, in which case you will carefully and objectively consider all constructive criticisms and possibly act accordingly. If, on the other hand, producing sequences is merely an ego inflating trip for you, then any adverse comments are likely to be dismissed as being irrelevant anyway. But if the latter were the case, then you probably wouldn't be reading this book in the first place, so I'll continue.)

Of course, the whole issue depends upon the motive for making the sequence in the first place. This is where success or failure hangs. If you are trying to arouse an emotion or reaction to a principle or set of circumstances then success is measured by the proportion of your audience who respond so. On the other hand, you may be using two projectors simply to avoid a blank screen between consecutive slides, and the music added to give another dimension. If your object is to show pretty slides under more favourable conditions than normal, then you are likely to succeed. You may not wish to go any further than this and there is no reason why you should. There are others, however, who, after the first sequence or two, realise the potential of this medium and wish not only to use it more fully, but to squeeze the last drop of versatility from it. Such people have a hobby or ambition which can last them for the rest of their lives.

The object of the sequence

There are many motives for making sequences, and an attempt to briefly summarise them all is likely to be inadequate. It is probably better to choose a limited number of motives and explore them a little.

Beauty. Although frowned upon by many, beauty for beauty's sake is a just reason for making a sequence. Why not explore, photographically, the features of a beautiful girl? Flowers, autumn tints, clouds, sunrises and sunsets have all been the subject of sequences, and, if treated well still can be most satisfying. Not Oscar winning material, maybe, but a useful contribution to a programme.

Education. Many of a captive audience would object to being educated if it were done under the pretext of a slide-tape programme. However, this can be done, without dissent, if the audience is primarily entertained and the education is, unknown to the audience, pushed in through the back door so to speak. I learned a good deal about Van Gogh, for example, by watching a sequence made by a friend of mine. The photographs were of the artist's paintings and the sound, in two parts, consisted of a song followed by a documentary. It was enjoyable to watch and at the same time it was educational. (Oh, if only education had been like that in my school days.)

Mystery/abstract. In some respects, this could be regarded as a technical exercise from the production standpoint, although the results can still be entertaining. It is possible to render fairly commonplace objects almost unrecognisable if small details are taken in close-up. Animal, vegetable and mineral matter can be made virtually abstract by this method and will keep your audience guessing, and hence hold their attention. A visit to a scrapyard can pay dividends, as can opening up a transistorised radio. Crumpled aluminium foil, a toothbrush, a bath sponge, a bundle of drinking straws can all be used with imaginative lighting in close-up. Couple the images with suitable (say electronic) music and the effect can be startling.

Humour. This can be the most taxing of themes, but, if successful, one of the most rewarding. To start a programme with such a sequence helps to break down any reserve that an audience may have; get them laughing and they are on your side from then on (provided that you do not bore them to death with an over-long sequence later). But what can you photograph that is funny? The limitations here are only those of your imagination.

Unlike the single photograph of a funny incident, which occurs only rarely, you can make the sequence funny. The secret lies in the script. How do you make a humorous script? If I could answer that one fully I would be earning a lot more money than I do now, but there are some general hints. Humour largely depends upon surprise, a predictable outcome loses the impact and, with it, much of the humorous potential. There are exceptions, of course, if the next line of the script is going to be obviously risqué, changing it or avoiding it altogether adds to the mirth. However, if hard rules for humour could be laid down, they would be self-defeating by being predictable.

Back to the photographic side of the humorous sequence. Technical expertise is low down on the list of requirements, imagination is near, if not at, the top of the list. I have joined in the laughter which was initiated by sequences whose subjects have been as varied as snapshots taken in Spain, clothes drying on a line, a (documentary on a) city and two cameras which were given characters. Remember, conferring characters on animals or inanimate objects can be a winner. There is little which is inherently funny in puppets, for example, but give them characters, and then what they do and say can make them soar in the popularity polls for entertainment.

Natural history. Regrettably, this subject, I believe, is more suited to the single photograph or cine rather than the slide-tape sequence. The enthusiast in this field usually seeks superb photographs, and is reluctant to show work which he feels is less than his best. This being the case, the work needed to obtain enough material for a sequence is enormous. However, when such a sequence is produced, it is usually of prize-winning quality. The best usually have a theme, such as the life of a "family" of animals, or the progress of natural colonisation of a new environment.

Documentary. The slide-tape sequence is well suited to the documentary. Many documentaries, and especially those with an historical content, seen on TV and the large screen these days involve photographs of paintings, drawings, etchings or other photographs. Apart from the panning and zooming effects obtained with the television or cine camera, much of the work involves still shots. Once again the limitations are imagination and skill, and there is no doubt that superb documentaries can be produced with this medium.

Advertising. In these days of rising costs, a great deal of advertising is being done by slide-tape. Production costs are much lower than film and TV and the medium is still novel enough to attract the attention of those who are used to the single projector or moving picture techniques.

Stories. This is very popular in France where many diaporamas (as they call slide-tape sequences there) have the telling of a story as the raison d'etre. The medium is certainly capable of achieving this aim as is witnessed by the number of story sequences produced. I have seen sequences on subjects varying from stories of the Middle Ages to space travel, and very good they were too! One word of warning, though. Like any other subject, it can be overdone. Sitting through a session of six or so sequences each one with a story or message can be a most taxing experience. I would say one story or message in a programme (or at the most two) is enough.

Which comes first, image or sound

Although this was touched upon in a previous chapter, further consideration is worthwhile.

Very early, one cold winter morning I was walking by a river bank (shamefully, without my camera!) with the shafts of sunlight cutting through the mist when I realised that this was a perfect opening for a sequence. With the novel twist which I thought up, it should be a winner! (Never mind what the novel twist was – think up your own ideas!) Obviously, the images come first.

That same night my wife and I were relaxing in a local bar when someone put on a record which neither of us had heard before. After a few seconds, we turned to each other and simultaneously realised that the other had had the same idea for a sequence. Obviously, the sounds come first. But this is rubbish! Both can't come first. Of course not, neither actually comes first. It was the idea which was prompted by the image or the sound, which started the mental ball rolling.

Consider the images coming first. Which images? You would not normally take a collection of slides and then simply put them together to form a sequence. You need a theme. What theme? Well, you could arrange them in order of ..., or again, instead, perhaps, they would be better if they followed the idea of.... Yes! The idea!

The same considerations apply to sound. I realise that many claim that music does not conjure up visual images. But does this matter? Anyone who has seen Walt Disney's Fantasia knows that sounds can conjure up images, and when put together, provide first class entertainment.

So, the idea is the beginning. At the outset, the skeleton of the idea gives you some identification of the images you need. More thought will fill in more detail, but it is better not to be too specific, because that limits the sounds available. So, next try to find some appropriate recorded music. It is unlikely that any single piece will meet your needs. More often than not it will be too long in duration.

The next question is whether to produce more slides to fill in the time, or shorten the music somehow. Both steps are far from ideal. Perhaps it is too soft at the wrong places or reaches a climax at an inappropriate part. The real solution is tape editing, taking different sounds from different sources and blending them all together to give exactly what you want. After all it is hardly likely that any composer will have produced a piece which is ideal for the idea which you have just had. So now you have progressed from your original sequence where the music, to a certain extent, dictated the images which you could use. To produce exactly what you want, you need to treat your sounds in the same way as your slides. That is, mix and blend them to control the sequence, not let it control you.

Visual/photographic effects in dual projection

The photographic tyro follows various rules of composition eg. in his early work: he puts the main point of interest at the intersection of the thirds; seeks the "C" shaped composition; arranges a balance between areas of different sizes and brightnesses (or colours) and so on. Such efforts are fine for single photographs as each is studied as an entity in its own right and without any specific relationship between it and any other photograph.

With dual projection, the situation is somewhat different. Apart from the first and last photograph in a sequence, each image is sandwiched in between two others; and, unless the change from one slide to the next is instantaneous, there is a period when both images are visible on the screen simultaneously. With skill, you can use the third image (consisting of the superimposition of two others) to advantage visually. For example, whereas the focal point of a photograph would rarely be tucked away in the corner of the frame with a single image, this could be desirable if the previous image consisted of a shadow in that corner. This can be especially effective if strong lines in the composition lead the eye to that area.

There are numerous such effects which, if used with discretion, can add enormously to the enjoyment of a sequence. However, like many effects, if used indiscriminately they become monotonous.

Changing focus. This is a very popular technique and is used frequently in film and television. It relies on there being two points of interest in the field of view, one near the camera and one further away. For slide-tape purposes, the method involves taking two photographs, at a wide aperture to reduce depth of field, one with the distant object in focus, the other with the near object focused. Cross-fading from one to the other gives the impression of depth on the screen and also changes the area of interst. The technique can be a powerful tool in skilled hands. For example an image of a man, in the distance, can show that he is miserable. Refocusing on the foreground (which was, in the previous image, almost unrecognisable) shows it to be a wire netting fence, indicating that the man is a prisoner of some sort, explaining his misery. Similarly, in a church a photograph of wrought iron gates show them dark against a background patchwork of brilliant diffused colour. Cross-fading to the next slide can show the colour to be a window and the iron work asa blurred dark mass.

Although it is possible to take such photographs with the camera hand-held, for precision in image position on the screen it is preferable to have the camera on a tripod. Even then, a certain amount of shift in image is unavoidable. The magnification of any image in the plane of the film changes as the lens is refocused. Also, out-of-focus image is larger than the in-focus image.

Highlights and shadows. The eye usually tends to concentrate on the highlights in a picture. So, when highlights and shadows occur in the same image the shadow areas offer a perfect opportunity to introduce a highlight and thereby lead the eye of the viewer across the frame. The eye can be led from side to side, or corner to corner, at will. Again, do not overdo this, or the audience may become tired of being forced to watch an optical version of a tennis match!

During cross fading when using this technique, you can take advantage of the double image halfway through the fade. As time goes by you will find yourself keeping an eye open for subjects which allow highlight to appear out of some specific region of the frame.

Ghosts. This technique has the potential for producing amusing effects, or can be used more seriously. People can be made to appear or disappear at will (without using a cassette of Highspeed Ectaplasm!) Take two shots with a tripod mounted camera; one with and one without the person in position, but otherwise identical. Cross-fading does the rest although the precision in mounting and projecting the transparencies is essential if the technique is to be effective.

Lightning. Sequences involving storms benefit from the occasional flash of lightning. The lightning can be pictured on monochrome film by leaving the shutter open during a thunderstorm. The flash will expose itself. After which the shutter can be closed, film rewound and shutter re-opened. It is essential that no light sources or light reflections appear in the field of view otherwise these will appear during the discharge on the screen (although with the necessary know-how, you can bleach out such highlights, with Farmer's reducer, on the negative or retouch out the highlights on the positive). To show such effects, project the slide of the scene without the lightning normally, at an appropriate place flash the lightning slide briefly onto the screen. Advance the magazine with the lightning slide while the normal scene remains on the screen. You can repeat the effect with a different image of lightning.

Time lapse effects. This has only limited succes with crossfading when compared with cine. Nevertheless, it has some value.

Photographs, say, of a flower bud opening, or even a chick hatching from an egg can be shown on the screen. Naturally, neither the camera nor the object being photographed should move during the series of exposures. With outdoor subjects it would be advantageous to form a wind barrier to prevent movement of the object between exposures.

Beware of using too many slides for this purpose, three or four slides are usually as many as can be accommodated before boredom sets in.

More successful is the long-term lapse effect, such as depicting a garden through the seasons. You can use, say, one slide for each month, provided that there is sufficient difference between one slide and the next. Your problem of course, is to ensure that the camera is in *exactly* the same position for each shot. A fraction of a degree of movement will make a large difference to the image framing on the screen and destroy the illusion.

Changing subjects. Two subjects of the same shape are photographed in such a way that each is the same size, and occupies the same position in the frame. Then cross-fading changes into the other, on the screen.

For this purpose a removable ground glass screen on a single lens reflex camera is almost invaluable. You can mark the precise positions of such objects lightly on the screen in pencil, temporarily. Failing that, a grid on the screen gives some assistance.

Changing time of day. This is really a modification of the time-lapse technique although effects can be rigged. Most people enjoy the occasional sequence involving the setting sun (but without the hackneyed "and so we say farewell to the beautiful ...").

Suppose you have a landscape with dominant foreground masses such as trees. You might like to finish with this image melting into a sunset at the same site and finally the light fading away. What a beautiful end to a sequence. The problem is that it will probably be another ten months before the sun sets in that part of the sky. If you wait that long it will probably be cloudy or worse. The answer is to create your own sunset. Any sunset in any part of the world will suffice (providing that the highlights are in the right position). Make a high contrast monochrome positive of the slide in question and mount this with the sunset. This can be used as the last slide or the sandwich can be copied for this purpose. One word of warning. Please make sure that the church spire and its weather cock are not in the picture, especially if this indicates that the sun is setting in the North!

Changing lighting conditions. You can produce some novel effects with a single subject if photographed in a fixed position with all variables fixed, except lighting. The technique is restricted largely, though not exclusively, to subjects photographed in the studio. You can try illuminating fairly commonplace subjects from different directions and/or with different colours of light. The effect, when such images are cross-faded, can add to the variety in a sequence or show. As with all techniques, it should not be overdone. More than two or three slides of a single subject, so treated, is likely to invite boredom, unless other variables such as change in viewpoint, are involved.

Mixed monochrome and colour. Although photography involving mixed monochrome and colour was used long ago (such as in the film "The Wizard of Oz") its usefulness is by no means exhausted. With slide-tape the initial impact is possibly greater especially if the two are mixed in the same slide. To see a monochrome image, especially a high contrast one, on the screen, build up section by section into full colour, is an experience which is hard to forget.

One method of achieving this effect is to start from a transparency. It is possible to work with 35mm but as much retouching and handling is involved, it is better to start with something larger (5" × 4" is a convenient size). As the process requires a number of separate images in perfect register, you need some form of registration system. If you intend a great deal of close register work, the punch-pin method cannot be beaten. Commercially available ones tend to be expensive, but a very serviceable one can be fabricated from an inexpensive office punch which simultaneously produces two circular holes in a sheet of paper. The pin register requires two round pieces of rod or dowel on which the perforations fit tightly. These should be fitted permanently to a suitable base

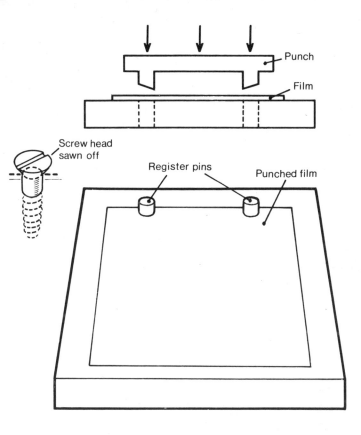

The punch-pin method of registration. Each piece of film is punched before exposure, placed on the base over the pins (the base having previously been taped to, say, the enlarger baseboard). This ensures that the separate images will be in exactly the same positions on each of the pieces of film.

If such images are to be re-copies on to 35mm colour film then the base carrying the pins is best made from a transparent or translucent plastic (e.g. acrylic) to enable the images to be illuminated from the reverse side of the base.

in such positions to enable the perforated pieces of film to be placed in exactly the same position with respect to each other.

For those who are unwilling to go to such lengths, the simple flip-flop technique is quite serviceable. Each image (up to 4) is taped along one side of the film area so that they can be hinged in exactly in register. The production of the monochrome images is governed by the effect desired, but it is usual to start from a monochrome negative produced by contact. From this negative you can make continuous tone or high contrast lith positives. These can act as masks in contact with the original transparency for the purpose of copying onto 35mm colour film.

Bas relief. Bas relief techniques are well known in monochrome, less so in colour and very rare in colour slide. The appearance on the screen is, to say the least, interesting. The effect should be restricted to rare occasions which need impact of this sort.

The method involves cross-fading a positive transparency and a negative transparency. The major problem is that the heavy colour mask built into colour negative materials overpowers all other colour effects. One solution is to make a negative by processing reversal film directly as a negative, ie. using standard negative processing techniques. You can have lots of fun with this one!

Synthetic colour. There are rare occasions when the use of synthetic colour on the screen can command the attention of the audience, but, like with all unusual techniques, its use must be treated with restraint.

Such processes as posterisation, solarisation and the use of contour materials all provide exciting effects on the screen, especially with images in register, but the details of such processes are, of course, beyond the scope of this book.

Presentation/photographic effects in dual projection

These are effects which are obtainable with two (or more) projectors.

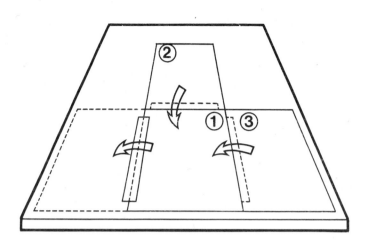

The 'flip-flop' method of image registration. Up to four images can be registered by this method. Image 1 is taped to the transparent base along its left edge. Image 2 is placed over image 1 and, when in perfect register, the top edge is taped. Image 2 is removed and the process is repeated with image 3, which has its right edge taped. A piece of fixed out film should be placed in the central position when 2 and 3 are being exposed, to allow for the thickness of image 1.

Horizontal/vertical changes. When slides are shown using a single projector there is little difficulty in mixing horizontal and vertical formats. With dual projection the situation is different. With anything other than an instantaneous change from one slide to another, there is a period of time when two images are on the screen simultaneously. This produces a cross-shaped image area which, for many people, is mildly irritating. Unless you have special reasons it is best to avoid mixing formats. Indeed, many slide-tape enthusiasts shoot exclusively in the horizontal format; at least when gathering material for sequences.

There are occasions, however, when mixed formats are deemed essential. When this is the case, it is best to keep all the verticals together as a separate section within a predominantly horizontal sequence. That minimises the number of changes in format. Probably the most obvious example is a sequence involving photographs of tall buildings.

How can the change be achieved satisfactorily? A number of methods are available.

One of the most effective is to break down the rectangular shape, photographically into a square shape. For example, when changing from horizontal to vertical the last horizontal shot of the section could be taken through an archway. Each of the edges of the photograph are thereby lost in shadow, in effect producing a square frame. A vertical image can now be projected without any adverse effect. The audience may not even be aware of the change.

A modification of the method above, used by some, is to produce the square shape physically. This can be done most easily by rotating one half of the slide mount through a right angle before the slide is assembled finally. The combination of a horizontal and vertical mask produces a square aperture.

Personally, I feel that this method looks rather contrived, but on the other hand, if the sequence is about a once-in-a-lifetime-holiday in a far distant country it is hardly worthwhile going back to obtain one photograph of an arch! (You could manufacture one in the darkroom though – but some would regard this as cheating!)

The change from horizontal to vertical can be made more acceptable to the eye if strong vertical lines occur in the horizontal frame. Such strong verticals can be provided by reeds, for example. If the same reeds appear in the next vertical image, and especially if the rate of fade is fast, then the cross shape during the change tends to be overshadowed by the psychological effect of the strong verticals.

When none of these methods is available, you can fall back on a simple snap change. Anyone in the audience who is irritated by the cross-shaped image, will, at least, be irritated for a shorter period of time!

The snap change. This is simply a very quick change from one projector to the other. It can be practically instantaneous with mechanical faders but is less effective with faders which vary the power to the projector bulb because of the finite period of time needed for bulbs to cool down and warm up.

To be effective the snap change should co-incide with "snaps" in the sound. Having snap changes in the middle of soft, flowing music gives an incongruous effect.

The jerk change. The jerk change can be very effective if it accompanies or is accompanied by appropriate sounds. When the music has a well defined march-like beat (and the visual material is appropriate) a change from one projector to the other in the form of a number of equally spaced jerks can be one of the best methods for producing audio-visual impact. A piece of music containing, for example, three rapid cymbal crashes simply cries out for a three jerk visual change.

The twinkle. This involves the rapid change forwards and backwards between two projectors. The technique can be used to give the effect of the movement of trees in the wind (two photographs from the same position are needed but with the branches in different positions, caused by wind). When music is very active and would otherwise require very rapid slide changes, the twinkle effect is a very useful technique to use (most projectors require something in the order of two to three seconds to change slides).

Non-alternate showing. When a single image is to be kept on the screen and is to be accompanied by a series of different images shown progressively over the main one, then non-alternate showing is required. One slide and magazine is left projecting the image while the superimposed images are provided by the other projector showing and advancing at the required rate.

The effect of two or three different lightning flashes over a fixed scene needs technique.

More ambitious sound

There can be little doubt that the best sounds for a sequence would be those composed and produced specially for the purpose. Professional film makers have been doing this for decades. In our humble way, we have to make do with sounds which have already been recorded or go out and record them live. Some of us may even be fortunate enough to have access to sound-making devices (no disrespect to musicians, I include electronic synthesisers). It is unlikely, though, that most of us will have sufficient skill to compose and produce the sort of sounds which we need. Our skill must be used to record and extract the recorded sounds which are wanted and put them together in such a way as to produce an integrated audio-visual experience.

Sound sources

It is amazing how much we take sounds for granted. Most people, if asked which of the senses they would be prepared to lose, if given the choice between sight and sound, tend to choose sound instantly. Then, on further consideration, retract this and say neither, when they realise how valuable their hearing is. Sounds are all around us, all the time, even in complete darkness. It is this very point which, paradoxically, often offers one of the major problems in recording sound. (There is too much noise around!)

Copyright

Before we consider recording sounds, there is the question of copyright to be considered. It is illegal to make (ie. not just play back) tape recordings of recorded sound be it from disc, tape, film etc., or from live performances without the prior consent of the owner of the copyright of the work in question. The fact that it is done frequently does not alter the legal position at all. It is better to be safe and obtain a license for your recording before you begin. Further details on copyright are to be found in Appendix 7.

Live recording

The human voice. Recording the human voice has problems all of its own. The first one is psychological. The effect is just like pointing your lens at some person and telling him to "smile please". The same sort of thing happens when you poke a microphone at someone's mouth and ask him to "say something" (this is the most sure way I know for silencing even the most avid chatterbox — at least temporarily). You probably froze-up yourself when you were at the receiving end of this exercise for the first time.

When the embarrassment has been overcome the next factor is *quality* of voice. At first we all tend to think that our voices are not bad in quality. Listening to our comments played back usually cures that illusion! The reason is that we do not hear ourselves as others hear us, a contribution to our hearing of our own voices is made directly through our own heads.

Whether your own taped voice is satisfactory or not depends upon the purpose for which you want it. If the object of the sequence is a documentary on say, a building or a village, and you possess the appropriate local accent then all is well. After a few trial runs, off you go recording merrily. If your voice doesn't have the required local flavour then you need someone with a voice which has. Also, get somebody who knows about the subject in question.

Perhaps you want the voice for the narrative of a serious

documentary of a general nature; or for a piece of poetry relating to the sequence. The requirements in this case may well be quite different. You may be looking for a more neutral voice. (Have you noticed how a person with a strong dialect is regarded, by many, as being not so well informed as someone with a neutral or "cultured accent"?) The only answer here is to employ someone with such a voice, perhaps someone whose voice has been trained. I know this all sounds very fussy and may involve a lot of trouble at the time, but it is better than regretting not having made the necessary effort when, later, the finished sequence is criticised on this point. Whether the atmosphere is to be recorded with the human voice is a matter of personal choice. Sounds of the sea, of machinery, or of traffic will all add to the atmosphere but they are out of your control so you may prefer to tape the voice separately and mix the atmosphere later when you can adjust the relative volumes to your liking.

Atmosphere. Those sounds which are always around us, but which we tend to ignore, are the same sounds which can make all the difference to the atmosphere conjured up during a sequence. The "caw-caw" of crows, the sound of the wind, the babbling of a stream can all be recorded and added to your sound track to impart that extra something which makes the show all that more interesting. Obviously, to record such sounds requires a portable recorder but often the quality need not be as high as would normally be required and, for that matter, mono sound is often adequate (two channel sound can be provided by using a mixer).

When recording sounds for atmosphere, monitoring, preferably through headphones is an important step. Our ears can be selective, our microphones cannot. I remember once recording the sound of fountains, the first try picked up the sounds from the pump-house beautifully. Attempt number two was accompanied not only by a groundsman who started his motor mower simultaneously with my pressing the "record" button but also by a plane, which, up to seconds earlier, had been lurking out of microphone range. There is no point in getting home before discovering about such problems, find out there and then, and take steps accordingly. With continuous sounds, such as fountains or the sound of the sea, it is not necessary to record for a duration equal to that of the sequence. Provided that there are no outstandingly identifiable sounds present in the recording, you can employ a tape-loop. This is simply a recording of a few seconds duration with the ends spliced together to form a continuous loop. This is played over and over again on an open reel deck. I have known people to support the excess loop on a series of beer bottles to avoid physical damage. (I suppose any other type of smooth bottle would do.) It is also possible to buy special reels that allow you to use tape of longer duration.

Animals. There is little more difficulty in recording tame animals (or those in captivity) than in recording atmosphere. The same problems exist. Wild animals are a little more difficult but those which can be approached to within a few dozen yards can usually be reproduced on tape with special microphones.

Concerts. By "concert" I mean any public appearance of any group of audible performers, musical or otherwise! Such sounds as applause will, of course, be included in the recording but it need not suffer from their inclusion. A trial run for the purposes of input control is desirable. If the sounds are spread over a large area such as with a large group or orchestra, you may need several microphones, strategically placed; and fed into the recorder through a mixer (page 141) which will permit the relative volumes from each of the microphones to be adjusted. A limiter switch, either on the recorder or on the mixer is a definite advantage as this will reduce any tendency to over-record, and hence distort, any sudden large increase in the volume of sound.

Naturally, you will seek permission to record before starting.

Recording from radio, tape and disc. Dubbing (the re-recording from tape) and recording from radio and disc presents few problems. You can feed the output directly into the recorder, or, if desired, into the amplifier and from there into the recorder. This method avoids the recording of unwanted external sounds. The use of the amplifier sometimes permits a certain amount of control such as filtering out of high-pitched scratchiness which may be present in records of less-than-perfect quality.

Recording from television through a microphone is not recommended unless a quality falling well short of hi-fi is acceptable.

There is one special use of recording from discs which is worthy of mention. A series of special sound effects records have been produced by the BBC specifically for recording purposes. The range of sounds available in this form is very wide and includes those which are natural as well as man-made. An important point concerning these is that copyright is waived, ie. it is not an offence to tape these sounds for public or private performance.

Mixing

Some years ago, while watching a sequence, I was puzzled by the sound track. Although the quality of the music was not good and the narration little better, it was not these that disturbed me. At frequent intervals, sounds came from the speakers which gave the impression that someone was shuffling a deck of dinner plates! After a while, I identified the cause. The sound track had been recorded by using a microphone not only to pick up the commentary but also the music from a record player, simultaneously with the slides being shown. Naturally, the microphone also picked up the sound of the slide change mechanism, with the result described above.

These are the sort of circumstances which make a mixer indispensable. With such a device sounds from various sources can be blended together at will. Some tape recorders have separate input controls for microphone and for "line" (ie. from another tape or from radio) and so simple mixing of a microphone input and material from another source can be achieved. However, anything which is more complex than this is best done by using a mixer. Some workers avoid using a mixer by recording a commentary on one channel and music etc. on the other. This technique, although avoiding the need for a mixer, gives a rather disconcerting effect on members of the audience, especially those near the speakers. Furthermore, it precludes the possibility of automatic projection with a conventional stereo (two channel) recorder.

Types of mixer

As with all manufactured objects, there is a range of mixers of different types and, inevitably, price brackets. Again, as the quality and versatility increases, so does the price. At the simpler and less expensive end of the range is the so-called passive mixer. This device enables sounds from various sources such as microphone, radio, tape and disc to be blended together. The passive mixer operates on the principle of attenuation, that is, each of the sound sources can be reduced (but not increased) in volume in order to produce the desired mixing effect. If any of the sound sources is of inadequate strength, then the others must be reduced accordingly, thereby decreasing the signal to noise ratio.

The more elaborate and versatile active mixer has inbuilt amplifiers to permit the balancing of different sources, as required, thereby facilitating the improvement of the signal to noise ratio.

Most of the more versatile mixers possess one or more panpots ie. panoramic potentiometers. These permit any single sound source to be distributed between two channels. Thus a commentary recorded in mono may be recorded on the left or right channels or distributed equally or unequally between the two as required.

Mixing down

There are occasions when stereophonic sound sources may be required to be mixed down to mono. Running automatic projectors with a two channel (stereo) recorder is a case in point. In some cases you can just take one channel, the left, for example without detriment; but with a large percentage of good quality stereo recordings, the channel separation is such that the overall quality will suffer. Perhaps a solo instrument of high pitch recorded on the left hand channel will be lost almost completely if the right hand channel only is recorded. The bass will suffer if the reverse is done. To get the full effect, both channels should be used.

When should mixing be done, at the outset when various sound sources are being taped prior to editing, or at the final dubbing stage of the edited tape when producing the master tape which will carry the mono sound and the control impulses for the projector? There are advantages in each, but the latter has the advantage that, if at any time in the future you acquire a three or four channel tape deck (so that stereo sound and control impulses may be recorded) it will only be necessary to dub the original edited tape again, rather that than start the recording project from the beginning.

Mixing in the commentary

Producing the voice over ie. adding the commentary to the background music or other sounds can be achieved in a number of ways. Using the commentator simultaneously with recording the main sound track is probably the least satisfactory method, being fraught with difficulties. One error by the commentator (it will probably happen near the end!) will mean starting the whole process again. Tempers are known to get frayed under such conditions!

The commentary is best recorded separately. Two or three attempts can be made as rarely are any two alike. The best fractions of each can be identified and incorporated, in correct order, into a master tape, leaving an interval of about four seconds between each section. When making the final tape, the pause control on the deck carrying the commentary can be used. This avoids recorded clicks and thumps which would otherwise occur if the deck were switched on and off. Copy the background sounds onto the final tape and about two seconds before the commentary is to be introduced, reduce the volume of the background and, simultaneously, release the pause control. After that section of the commentary has been recorded, let a two second interval pass before engaging the pause control, and increasing the volume of the background sound to its former level. (It is under such conditions that the advantages of slider controls over knobs is realised, especially if the former have preset indicators). Repeat the process for each section of the commentary.

Scriptwriting

Before any knowledgeable person screams in anguish at scriptwriting being relegated to a section-of-a-chapter-of-a-book let me hasten to add that scriptwriting is, in itself, a profession. Visit any really comprehensive library and you will find whole books devoted to this subject alone. What chance is there of doing justice to this subject in such a small section? Practically none at all, but nevertheless let me try to give you some guidelines, at least.

If it comes to that, taking photographs is a profession, and so is sound recording; but we are trying to be clever. We are doing the whole lot ourselves. And why not? Some first class work is being done by amateurs. They, unlike professionals, can spend a full year, or more, getting one sequence exactly how they want it. Professionals have deadlines to meet, economics to consider and (often irate) clients to satisfy. Lucky amateur!

Obviously, sequences of pretty images accompanied by pleasant sounds do not need a script. Nor do sequences created by fitting images to a piece of poetry or lyrics of a song. A script is needed when images and music together are not enough. If the sequence involves the communication of information, then more often than not images music and words are needed. (Research tends to indicate that about 10 per cent of what we hear is retained, twice that amount of visual information, but when information is received through both eyes and ears, over 60 per cent is retained).

Let us get back to the scriptwriting. The first thing to decide is the object of the sequence and which aspect of the medium is best suited to this aim. The scriptwriter for an advertising sequence may have a free rein, as appropriate images will be taken later to illustrate the points made. Conversely, the script for a holiday sequence is restricted by the limited nature and scope of the visual material already available. Ideally, you should work from an overall plan for both visual and aural material. Only in this way can you avoid that all too common frustration of wishing that you had taken another shot from this angle, or nearer, or at another time of day.

If there is one common pitfall in this medium, it is to use a technique, idea etc to excess, and the script is no exception. While the script is important, it should not dominate. Remember, this is an audio-*visual* medium. Whatever you do, do not let the commentary drone on and on incessantly. If you do, your audience will suffer from aural indigestion and will close their ears.

Your sounds should supplement the images so allow plenty of time for them and the music to entertain your audience in between sections of the commentary.

Stating the obvious is another error to avoid. With an image of the Eiffel Tower on the screen, the worst thing you could do is to say "this is the Eiffel Tower in Paris". Everybody knows what it is (unless you are entertaining visitors from Mars on their first visit to Earth!). It is far better to say something like "this graceful construction of iron was built for the great Paris Exhibition of 1889". In this way you use each part of the medium to its best advantage. Of course it means that you have to do a bit of homework on the visual material which you have, but it is well worth the effort and it can be done at your leisure, usually at your local library.

Let us take an example of scriptwriting for a series of slides which are going to be a part of a sequence. We can look at two different scripts. I am not going to claim that one is "right" and the other "wrong". They are simply different. One, I believe, is better than the other, in as much as it entertains, and informs the audience more effectively. This is not to say that the better one cannot be improved. I am sure that it can. It is simply an example of the direction in which to go. Suppose we have five slides of Stonehenge.

Number 1 is a distant shot, the second a single vertical stone with a man near it. The third is a photograph taken with a deeply coloured filter over the lens, to give impact. The fourth slide is of a sunset behind a silhouette of Stonehenge (this could be a sandwich) and the last shot is at night and includes the moon.

IMAGE	SCRIPT
1) Long shot of Stonehenge.	"This is Stonehenge from a distance ...
2) Close up of single stone with figure.	and here is Uncle Fred standing next to one of the stones".
3) Photograph taken with deeply coloured filter over lens.	"This is a nice shot. I used the filter that Mable bought me last Christmas".
4) Silhouette and setting sun.	"There is Stonehenge at sunset".
5) Slide taken at night, including the moon.	"I waited ages for the moon to come up for this one".

There is nothing in the commentary here which the audience does not know already from the photographs, or if there is, it is irrelevant anyway. This commentary is more or less superfluous.

Now let us try to incorporate some factual evidence, obtained by some research on the subject and flavour it with a little imagination.

IMAGE	SCRIPT	COMMENTS
1) Long shot of Stonehenge.	"There are few, if any, who cannot identify this historic landmark on Salisbury Plain. Many attribute the construction of Stonehenge to the Druids but contemporary thought dates it much earlier. The enormous size of the individual stones cannot be appreciated from this viewpoint so let us move closer –	(You have not insulted the intelligence of your audience by telling them what they already know. There is information here of which they were probably unaware).

Fade to next slide. |
| 2) Close up with figure. | – Now we can appreciate its size better, as a visitor lends scale to the scene ... (pause) ... There is some controversy about how such large stones were brought here from Wales, some 200 miles away so it is little wonder that some early | |

146

writers claim that
it was all done by
the Art of Merlin
the Magician!

3) Photograph taken
with deeply
coloured filter
over the lens.

Quick fade to
filtered shot.
Background
sounds change to
something suit-
ably "weird" or
"magical".

4) Silhouette and
setting sun.

As night falls, the
atmosphere
changes, we
could be among
those primitive
people, gathering
here for a cere-
mony. Or, per-
haps, the more
scientifically
minded are
absorbed by the
skill of those
ancient builders
who placed their
stones to make a
religious calendar
to predict, not
only the Sun's
movements ...

Slow fade to
sunset.

5) Slide taken at
night including
the moon.

but also those of
the moon".

medium fade to
slide with moon.

The script can continue in a similar vein to give your first draft. However, before serious taping commences it will pay to perform a trial run. The script can be read out or the recorded script played back while you project the slides. Pay particular attention to whether the slides are changing too rapidly or too slowly and make amendments either to the script, or to the number of slides in the sequence. For example, the section of script above may, depending on the narrator, last for one minute. This will correspond to five slides only. Are those five interesting enough to be shown for an average of twelve seconds each? Possibly not, but you can introduce more slides if you do not want to shorten the script (remember the advice on page 88 about taking plenty of photographs?).

With some documentaries, the need to write much of the script can be avoided by recording an interview/conversation with the person involved in the subject of the sequence. If this is done "on site", any background sounds will add to the overall effect. If your interviewee has a strong regional accent, so much the better, as long as his diction is clear. Obviously, the use of a "cultured" voice for this purpose would be synthetic. The trained voice is best reserved for use in the more general aspects of the subject.

Organising and Showing a Programme

Now you have a number of sequences and, naturally, want to show them. It seems hardly worthwhile showing a single sequence of say, fifteen minutes, for it would take twice that length of time to set up our equipment, and almost as long again to put it away. Far better to make a programme of sequences lasting, perhaps, an hour.

Putting a programme together

How do we go about putting the programme together? Certainly not in an arbitrary manner. A carefully considered order of showing pays dividends in terms of increased enjoyment for all present. Naturally you will not put a weak sequence at the end of a programme. What is wanted is to leave the audience clammering for more. OK then, put your best sequence at the end. Similarly you want to interest the audience from the outset so that an appealing opening sequence is called for. Something with a humorous flavour would be ideal. Get them laughing at the start and you have won half the battle, they will be on your side from then on.

Perhaps you do not have a funny sequence. What then? At all costs, command their attention. Why not make a short introduction? This could take many forms. You could introduce yourself, and any helpers, on the screen. Perhaps, poke gentle fun at all the effort that has been involved in putting the show together. Another possibility is to use abstract type images and appropriate sounds as an introduction. People often become curious about images which they can't quite identify. There are lots of possibilities; it's your show so its up to you to instil into the programme something of your own character.

With the beginning and end organised, the important task now is to arrange what goes in between. When you have a large number of sequences, you have the obvious advantage of being able to select sequences and, more important, reject sequences which do not fit into the programme. (The principle is the same as choosing slides for inclusion in a sequence.) An important consideration at this point is variety. Ensure that the subject matter, mood, tempo etc. is varied in the programme and you will go a long way to keeping your audience happy. If you have three flower sequences, reject two of them. You can put them in another programme. Follow a lengthy sequence with a short one or a 'heavy' sequence with something 'light' and trivial.

It is most important to maintain variety in this way. Anyone who has sat through a series of five or six lengthy sequences, each with a 'message' and each demanding a fair amount of concentration, knows what a taxing experience this can be! Your audience wants to be entertained, not indoctrinated! By all means make your personal comment about your pet hate, be it pollution or Man's inhumanity, but for goodness sake (no, for your audience's sake) brighten the scene with something lighthearted or something such as beauty for its own sake. Remember the slide tape medium does not have the same versatility as film or television so don't attempt to compete with them. It is all too easy, for example, to produce a programme which lasts too long.

A factor which can affect the duration of a programme is the tape. If a cassette is being used then an uninterrupted showing of more than one hour is not possible. However, an audience rarely objects to a short interval in a one hour programme so that this need not be taken as a major disadvantage.

Magazine changes. It is inevitable that, during a programme of something in the order of one hour, you will need magazine changes. The larger the capacity of the rotary magazine has a self-evident advantage here. With an average showing rate of about eight slides per minute, a one hour programme will require a minimum of 3 changes on each projector with magazines of capacity 80 slides, whereas 5 and 7 changes each will be required with straight magazines holding 50 and 36 slides respectively. Each change in magazine is a potential source of trouble. What happens to those carefully arranged fades if the order of slides is changed by putting a magazine in the wrong projector? The audience will be even more aware of something amiss if a magazine change is made too slowly and loss of synchronisation between image and sound occurs. This can be especially serious with automatic projection. No, a little care in preparation can avoid such problems – and the attendant risk of the projectionist suffering a nervous breakdown during the show!

To facilitate magazine changes each full sequence should be scrutinised at the points where such changes are imminent. Look for a slide which stays on the screen for, perhaps, ten seconds. That is the point where the magazine change for the other projector should take place, even if that magazine is only three quarters full. There is little point in loading each magazine to full capacity if it leaves you with only four or five seconds to make a change. One friend of mine has built a microswitch into each of his projectors which causes a small bulb to light when a magazine change is to be made.

Another useful idea is to code your magazines so that there can be no mistake as to which projector it belongs. Remember, you will be operating almost in darkness.

Preparing to give a show

The same care must be taken in preparation for showing as in putting the programme together. Minute details must be taken into account, for example, it could be disastrous if you arrive at the appointed venue and find that you have forgotten the leads from the tape deck to the amplifier! No, nothing should be left to chance.

Ideally, if you are showing a programme at an unknown venue, it is advisable to make an initial reconnaissance, time, distance and other factors permitting. Investigation into possible positions of screen and projectors is needed. The latter depends on a power supply. The type and number of sockets for your power need investigating. You may need a universal plug (to fit any type of socket) and also extension leads for power. You are likely, too, to want a multiple socket. Is it a lecture-room with a stepped floor? Is your projector stand high enough to avoid the need to tilt projectors upwards? Are your projection lenses adequate for the image size and projection distance? These are the sort of questions which need consideration if the show is to run smoothly and on time. If distance precludes an initial visit, a letter can answer some of the questions but it will be as well if you prepare for the worst. Take everything that might be needed.

For this purpose a checklist can be invaluable, not only for the outward journey, but also to ensure that you bring all your precious gear back. A typical sort of checklist is to be found in Appendix 11. It is merely a suggestion and can be modified to suit individual requirements. However, a few comments explaining the inclusion of some of the items may not come amiss.

The inclusion of a synchroniser, buzz-box, ear piece, cue sheet and stop clock seems pointless if you use automatic projection. But what happens if your electronic fader malfunctions? (I have heard of a case where each power surge caused by the operation of a lift caused a slide change to take place.) Play for safety and have an alternative method for timing. The only time you will really need that alternative method is the time when you have not made provision for it. The headphones facilitate the locating of a specific point on a tape without the need to play back through the speakers and the microphone is useful, in a large hall, if you prefer to use your audio gear as a public address system rather than talk to the audience directly. The adhesive tape helps to keep speaker leads on the floor and prevents clumsy feet from pulling amplifiers from their appointed positions. Finally, that recorded background music will help to soothe those inevitable early arrivals while you are busy with the serious matters of setting up the rest of your equipment.

Packing for convenience

You can avoid some of the more tedious aspects of going through the checklist by keeping smaller items such as leads, bulbs, and fuses in a single package. A block of foam rubber or expanded polystyrene with individual recesses cut out for each item will enable you to check the presence of a large number of small items, quite literally, at a single glance.

Setting up for showing

The first consideration is that all the audience must be able to see the screen. This at once poses a problem. If the screen is at a high level, the projectors need to be raised so that their optical axes are on a level with the centre of the screen. This may not be possible, so they may need tilting upwards, thereby producing a 'keystone' effect. This can be cured by tilting the top of the screen towards the audience, if this is possible. Such problems become smaller as the projection distance is increased but this involves the use of projection lenses of longer focal length (and higher cost!) and also longer leads from the amplifier to the speakers.

Registration of images on the screen poses few problems if electrical/electronic methods of fading are involved. When mechanical faders are used, you need to take a little more care in arranging the optical axes of the projection lenses to pass through the centres of the apertures of the diaphragms. The use of registration slides does the rest.

When everything has been set up, a preliminary short run of a part of the introduction or the first sequence is all that is needed to ensure that all is operating satisfactorily.

Possible problems

One of the biggest disasters that can happen during a show is for a magazine to fall and spill its contents over the floor. Be careful to place magazines in a safe position in such a way that they can neither fall nor be put in the wrong projector. If, for any reason, a single slide is lost, there is little point in rummaging around in the darkness to try to find it. A simpler solution is to choose another slide from the same magazine, place it in the vacant slot and replace it after it has been shown. Provided that the slide has the same basic subject material as the one which went astray, the audience will not be overtly aware that anything is wrong. Placing the slide in reverse will minimise the ease of identification (but do not do this if there are lettering or clocks etc. in the picture!).

Another disturbing occurrence is a projector bulb blowing. This sort of thing usually happens when the power is switched on. Obviously, a number of spare bulbs will be insurance against this ruining a show. Three spares are not too many to hold in readiness.

Breaks in magnetic tape are uncommon but some may feel that a splicing kit is so invaluable on such occasions that it is worth including. A copy of the master tape of the programme is worth having also. Ninety nine percent of the time it will not be needed, but. . . .

What Went Wrong?

There is nothing more satisfying than one of your audience asking "when will you come and give us another show?" That is the most sincere form of praise and makes all the preparation worthwhile. Less frequent, but probably more important, is the member of the audience who waits for a quiet moment and after an introductory apology for not wishing to offend, offers constructive criticism such as "I think that the sequence about ... went on for rather too long" or "I wasn't sure of the point you were trying to make". This is the sort of feedback which can be invaluable if you wish to improve your work.

Oddly enough, there are only a few pitfalls waiting for the maker of slide-tape sequences but many of the lesser experienced seem to be invariably attracted to them. Being forewarned should offer predictable advantages and so, excluding purely technical boo boos (as described in the last Chapter), we can examine some of the more common causes of audience discontent.

Sequence duration

One of the most unforgiveable sins which you can commit against your audience is to bore them by having a sequence which is too long. It is better for your audience to want a sequence to last for five minutes longer rather than one minute less. This is a remarkably common fault, and there are numerous reasons for this. One is a matter of image selection. It is not uncommon to expose four or five times as much film as is needed for a particular subject and it is natural to try to avoid excessive wastage. However, it is essential to be ruthless in choice of transparencies for inclusion in any sequence. Not only is brevity the soul of wit, but it is also the essence of a successful sequence. With film and television, advantage can be taken of a story within a story; a changing expression on a face; a tiny *moving* object can command attention until it grows into an identifiable form; all of these are standard ploys. With slide-tape, this cannot be done. Such versatility is not inherent in the medium. "Make your point, then get on with the next one" is probably the safest idea to adopt. Perhaps part of the reason for this lies in contemporary trends. The description of minute detail present in novels by Dickens reflected the trend in those days. Probably today, such detail would condemn a new novel to failure.

Another contributory cause of sequences being too long is the problem associated with the sound. If, for example, parts of one or two pieces of music are to be used as background sound, there are a limited number of places where one passage can be terminated and smoothly blended into the next. Stopping a piece in the middle of a bar has a jarring effect on an audience and, except for some special purpose, is best avoided. Your audience will not be interested in excuses for a half-hour sequence simply because a chosen piece of music lasts that long!

The least valid cause for excessive duration is the assumption that what a sequence lacks in quality can be made up for in quantity. This argument has never been justifiable. If the object of the sequence can be achieved with twenty-five slides, then don't use thirty-five. Aim for quality!

Occasionally a subject is chosen which has the potential of lasting well over an hour. In my opinion, even the most skilful slide-tape producer would have great difficulty in holding the full attention of the audience for this period of time. It may well be that you have taken several hundred transparencies while on holiday in some exotic faraway part of the world but it would be better to make several short sequences of different facets of the subject rather than attempt a mammoth epic. When you watch sequences in the future, time each one and make a mental note of those which hold your full attention for the entire duration of the showing. It will be odds on that the

time will not exceed twenty to thirty minutes!
(Incidentally the same arguments apply to individual slides. Unless a slide is of exceptional merit, there is no excuse whatsoever for showing it – or more than one – for periods of twenty to thirty seconds.)

Overdoing the single theme

This is closely allied to the previous problem in that too many slides are used to say too little. It is very tempting to try to use a specific technique or idea to its very limits; but it is essential to be critical about the outcome. Take, for example, the time lapse technique used to show a flower opening. With cine, a period of twenty-four hours or even more can be condensed into a few seconds and this is perfectly acceptable as a part of a larger theme. The effect of continuous movement cannot be obtained as satisfactorily with slide-tape and if a large number of slides are used to attempt to counteract this, the overall duration will become excessive. (Remember the limited speed with which a projector can change slides!) Generally, three or four slides of a limited subject such as this is the maximum that can be used without the effect becoming ponderous.

There is a remedy, however. Photographs of limited subjects can be dealt with satisfactorily whether they are church windows, pieces of seaweed or simple snapshots. What is needed is a theme or story. In this case words, spoken or sung, are necessary. It is admitted that the object of the sequence may be changed from objects of beauty for their own sake, to something different. Whether this is acceptable or not depends entirely on you.

The subtle approach

We are getting into the realms of very personal issues here. What is subtle to one may be a bald statement to another. I believe that subtlety must be used sparingly with this medium. With the printed word the reader has time to "read between the lines", going over the same passage many times if necessary. Even the simple statement "that is a man" can mean different things according to which word carries the emphasis (or, for that matter whether it is followed by a question mark). With slide-tape the audience must appreciate the point at the first showing, or miss it. With this medium we cannot take advantage of a change in expression, lip-synchronisation or a natural progression of very subtle changes as can be used in film and television. Our images, by and large, are discreet, separate, packages of information and therefore subtlety is probably best left to the sound, if it is to be used at all.

Another source of potential confusion for the audience lies in the use of an enigmatic script or piece of verse etc. as the theme for a sequence. If such a script needs to be read several times before its meaning begins to be appreciated, then it is too complex to be taken in during a single showing. If, in addition, it is accompanied by abstract type images, for example, then the whole audio-visual experience will be based on confusion. Some of your audience may not object to this, others will try to 'understand' it, and there lies your problem. Do this sort of thing if you must, but expect it to initiate some controversy.

The literal approach

This is the opposite problem to the last one. It can result if images are taken to fit an existing script or lyric, especially if it is metaphoric in content. To take an extreme case, using Blake's "Give me my bow of burning gold, give me my arrows of desire.... Till we have built Jerusalem....", a literal illustration could involve photographs of bows and arrows, clouds and bricklayers' trowels. Fine, for a parody, or for a demonstration of how not to make sequences. Such images would be disconnected and predictable. No thank you, find another scriptwriter – or at least another script.

Mis-matching of images and sounds

Again, this deals with aesthetic factors and inviolable rules are difficult, if not, impossible to make. For example, a preponderance of bright red and orange images are not normally compatible with soft, gentle music. Sunrises and sunsets could be exceptions though. This is one of those things where you simply have to play it by ear (and eye!).
The experienced can break the "rules" purposely. The blending of incongruous pairs of images and sounds can produce hilarious effects, if done with skill!

Regularity of changes/fades

Predictability can be the death-knell of the slide-tape sequence. Examine what it is that makes successful entertainment and variety will come high up on the list. Whilst it is not always possible to introduce variety within the same sequence we should, at least, endeavour to avoid monotony. Nothing can be worse than constantly changing slides at the same rate over a period of time. Our organic subconscious clock will be quick to identify a constant $7\frac{1}{2}$ second interval between slide changes and before long the feet will tap out a steady beat of eight to the minute. No, avoid this one like the plague.
Similarly a constant rate of fade soon begins to pall with an audience. A mixture of snaps, twinkles, fast and slow fade rates helps to avoid monotony and hence increase the chances of entertaining the audience successfully.
These are among the reasons for the comments in an earlier chapter claiming that automatic fading devices with infinitely variable fade and slide change rates are to be preferred over those with which such choices are limited. At the outset, the use of an automatic fader with fixed fade or change rates is quite thrilling. After a time, however, more versatility is almost invariably sought.
Well, those are the more common causes of dissatisfaction with the slide-tape sequence. Avoiding them will not

guarantee automatic success, but will virtually ensure that you exclude certain failure!

Watch as many sequences as you can. Analyse what makes them fail, or succeed. After a while you will develop one of those indefinable "feelings" for the medium, and once you have done that, you will be hooked. It will be difficult to tell whether you have dominated the medium, or the reverse!

Where Do We Go From Here?

It all depends on where your interests lie, of course. Perhaps you are quite happy to continue to produce slide-tape sequences of increasing quality and impact. And why not? However, unless I am much mistaken, for those who have been really bitten by the slide-tape bug and whose self-imposed standards are rising constantly there will come an occasion (if it hasn't already arrived!) when that little bit extra, over and above two projectors and mono sound, will be sought. Do not misunderstand me, though. For the average enthusiast, there is enough versatility with two projectors and mono sound to give a lifetime of enjoyment. This chapter is concerned not with improving the existing medium but with extending its versatility.

Stereophonic sound and automatic projection

For those who are happy with manually operated fading techniques, the use of stereophonic sound presents few problems. The situation is quite different if automatic projection is involved. Whatever type of fading device is used, it will require a track on the tape on which coded signals will need to be recorded. With the standard type of stereo recorder this means that only one track is available for sound.

Many shrug aside stereo sound as a needless complication and extravagence where slide-tape is concerned. They claim that the images command more attention than the sound. In this, they are quite correct, but the argument does not end there. Few people who compare mono with stereo sound would care to deny the superiority of the latter, and these days almost all recorded material on the general market is intended for stereo playback. Not only is there a more 'full' sound with stereo, but extra qualities are present which are noticeably absent when a mono sound is played through two speakers. Now, while it is admitted that the majority of slide-tape sequences can be accompanied quite adequately by mono sound, there are occasions when stereophonic accompaniment gives an overall effect which is of considerable value. The 'panning' of a sound from one side to the other emphasising visual movement across the screen is an obvious example; split screen effects described later are another.

What are the requirements for stereo sound and automatic projection? Obviously, three channels are essential, two for the sound and the other for the control signals. Why not use two tape recorders, one for the sound and the other for the control of the projectors? There are at least two reasons. One is that considerable difficulty will be experienced in starting all the recorded signals at precisely the same time. The recorders may be able to be started simultaneously by electrical or even mechanical methods, but it is the appropriate part of each of the two tapes which must pass the playback heads simultaneously.

Even if, by much manipulation, such requirements are fulfilled to start the programme, the second problem arises. This is the improbability that both tape decks will run at the same relative speeds throughout the programme. So the two decks are likely to be out of synchronisation after any prolonged period of running. No, the only way to ensure that sound and control tracks are perfectly synchronised is to have them on the same tape, but before there are howls of anguish at the price of four channel tape machines, let me add that there are several ways of achieving this goal. Let us consider each in turn.

The four channel (quadraphonic) tape deck

This is the most obvious answer to the problem of storing three channels for information on one tape. The advantages of the open reel machine are retained; but in addition further advantages accrue from having four independent channels. You can record projector control impulses on one of the two channels which are non-audio. The other can carry recorded instructions or pulses, as a safety measure, to guide the projectionist for manual operation, in the event of the fading device malfunctioning on playback. 'Simul-synch' (simultaneous synchronisation) facilities built into some models such as the TEAC 3340S permit recording from several sources to be synchronised with ease.

The improved sound quality obtained by the use of tape speeds of 15 ips, available on some models, is also a factor worth bearing in mind. If you envisage the use of three or four projectors or multiscreen presentation, then the four channel deck is the obvious choice for control purposes.

There is one disadvantage to the versatile open reel four channel deck. It is big and heavy. Such machines must be heavy in order to contain the components necessary for their function (remember they need to do twice as much as the average stereo machine). Transportation is, therefore, more troublesome than with their stereo counterparts. Among the slide-tape fraternity, the equipment needed to be carried about for shows is often referred to as the 'hernia kit'. The incorporation of a heavy quadraphonic deck adds to the efficiency in that respect!

Before leaving this section it is worth mentioning a comparatively recent arrival on this scene. This is the four channel cassette deck. The reactions, when I have mentioned this topic to some people, have varied from howls of mirth to definite signs that they think that I should be certified! "You can't store four channels of information on tape which is only $\frac{1}{8}$ inch wide", is the first objection, until I point out that that is exactly what is done on the standard stereo cassette (two channels in each of two directions). One of the first of such decks is the NEAL model 140. With Dolby noise reduction on the audio channels, its performance, soundwise, cannot be faulted, except perhaps by the most fanatical audiophile.

The three channel tape deck

As far as I am aware, only one manufacturer, viz. Neal, produces cassette tape decks specifically for the slide-tape market. The "AV" models (102 and 103) have facilities of a standard stereo deck (with Dolby N.R.) and, in addition, provision for recording on an extra channel for projector control purposes. Such models are considerably less expensive than open reel quadraphonic decks and, of course, can double up for normal domestic hi-fi purposes.

The two channel (stereo) tape deck

For the purposes of stereo sound and automatic projection only the open reel machine offers possibilities for modification.

The most obvious means of recording a third channel of information is to employ the external synchroniser alongside the deck. The electronic circuitry of this may need to be designed to ensure compatability with the fader. Once this has been done there should be no reason why projector control cannot be achieved, at least in theory. In practice, however, such steps seem to give inconsistent, and hence, unsatisfactory, performances. The problem appears to be one of lack of accuracy in positioning of tape and playback head. A keen D.I.Y. fan may find a challenge here!

One method which has been shown to work satisfactorily is to mount an extra record/playback head on the panel of the tape deck itself. This has been discussed briefly on page 66.

Split screen techniques

There is no reason why the entire area of the standard 36 × 24 mm aperture of slide mounts must be used. For certain purposes this area can be divided and fractions of it used to advantage. For example, emphasis can be placed on a specific area of an image by increasing or decreasing the visible regions around it, by masking on successive slides. As with all techniques which vary from the norm, this one should be used sparingly otherwise the audience are likely to become a little disturbed. The use of 'gimmicks' like this may demonstrate your technical versatility but will not necessarily endear your audience.

Perhaps one of the safest ways to use the split screen technique is to make a full sequence based on this principle. I recall, with a chuckle, a short introductory sequence consisting of a dialogue between a judge at a slide-tape festival and one of the competitors. Each occupied half of the frame, one on the left and the other on the right. Both parts were played by the same person with (exaggerated) suitable attire in each case. Stereo sound added to the effect. The overall effect was most enjoyable. The attention varied from the left side to the right and there was no jarring effect on the audience (as can happen if one misses a change of images on one part of the screen while watching another).

It is worth pointing out that splitting a screen into three separate areas requires considerable skill if it is to be done successfully. Similarly, splitting images on the screen horizontally is rarely successful. Not only are the shapes of such areas rather awkward but also the technique appears unnatural.

Three projectors on one screen

Occasionally it is found that a specific visual effect cannot be achieved satisfactorily with only two projectors. Perhaps a snappy introduction to a sequence requires titles or other images to be superimposed over a background twinkle effect. In such cases there is no alternative but to employ a third projector. Generally, the third projector need not be of the same quality or even same type as the prime pair. The major question is how can the third projector be employed conveniently? Obviously, no problem exists if a four channel recorder and a second electronic fading device is available but if not, all is not lost. Here are some suggestions.

Manual control. When the main projectors are on automatic

control the third projector can be operated manually and, provided that timing is not too critical cueing can be by memory or cuesheet. Even a simple projector with a push-pull type of slide-change can be brought into use for this purpose as long as a large number of slides is not involved or the rate of slide change is not too fast. A blank slide is needed at both the beginning and end of such a series. The possibility of projector movement during the sequence is always present when manual operation is involved; but with title slides having pale lettering on a black background, this need not cause concern, unless precision in register is essential.

If you want the third image on the screen to appear and/or disappear gradually, then you need a fading device. It is also necessary with projectors which do not have a shutter to eliminate light on the screen during slide changes.

Automatic control

As long as fading of the third projector is not required then automatic control can be effected by use of a synchroniser. With open reel recorders an external synchroniser can be

THE EFFECT WHEN ONE COLOUR IS PROJECTED
OVER ANOTHER
(with either two or three projectors)

Colour 1	Colour 2	Effect produced*
Red	Blue	Magenta
Red	Green	Yellow
Blue	Green	Cyan
Red	Cyan	White
Green	Magenta	White
Blue	Yellow	White

* These effects assume each light source is of the same intensity.

used with no restriction. Internal synchronisers may be used on both open reel and cassette recorders but the pulse type synchroniser will necessitate losing one sound channel, ie. mono sound only will be available. Stereo sound can be retained if a pause type synchroniser is employed.

General considerations when using a third projector

The colour of images from a third projector will not necessarily be the same as when the same slide is used with a single projector. There are two reasons for this, one being physiological and the other psychological. The visual effect when two different colours are projected onto a screen can be predicted with fair accuracy. The three primary light colours of red, green and blue mix to produce the secondaries yellow, magenta and cyan. Mixing all three in appropriate intensities gives rise to the appearance of white. Consequently any pair of complimentary colours, when projected onto a screen will produce white, eg. red and cyan, green and magenta, or blue and yellow. Therefore, in order to obtain, for example, white lettering from the third projector, over a series of cross-fading images of a predominantly blue colour then it would be of advantage to use the letters in yellow, rather than white.

All this theory works in practice under a limited set of conditions. However, the psychological effects also come into play and these cannot always be predicted. The visual appearance of a colour depends very much on surroundings. Hence a given green colour will look much darker if surrounded by yellow and much lighter if surrounded by blue. It will be impossible, visually, to tell whether the greens match each other without masking off the surrounding areas. (The effect can also be observed with grey on a black and a white background.) The effect is further complicated by the fact that this illusion varies with varying relative sizes of colour area and surrounding area.

In conclusion, it is evident that after an initial theoretical consideration of the means of producing a specific colour effect, the final effect will need to be achieved by trial-and-error methods.

Back projection

Under conventional conditions of a slide-tape show the projectionist and his equipment are separated from the screen. This necessitates long leads from the amplifier to the speaker (with the attendant risk of all types of damage). A set of projection lenses of different focal lengths may also be needed for different sizes of room and projection distances. Such complications can be eliminated by using back projection. The principle involves projecting the images onto the back of a translucent screen which is placed between the projectors and the audience. Projection lenses of short focal length are employed to minimise the projection distance. For direct projection, the slides are placed in the magazines the 'wrong' way round. Otherwise, a 45° mirror may be used with each projector to reverse the images on the screen and also economise on the distance required behind the screen. Special screens are available for back projection although thin white linen stretched taut over a framework is quite suitable. At first it seems as if back projection is the perfect answer to all projection problems. But there must be a snag somewhere. Of course there is! The distortion of the rectangular image shape is more pronounced at short projection distances. Consequently, keystone or trapezium distortion will be much more serious than with conventional projection. The answer is to conceal the lack of register of the perimeters of each image by using a screen with a broad black border to mask off such regions.

Two other methods are also available. One is to use slide mounts with specially distorted masks so that, on projection, at the predetermined angle, the image on the screen appears perfectly rectangular. However, images with straight lines running parallel to the edges of the transparency (eg. buildings) must be avoided or the visual effect will look most odd. The other method is to use those mirrors mentioned above. If

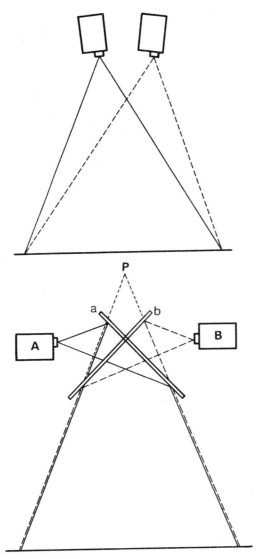

Back projection. Above, with short projection distances, projector spacing can give considerable image distortion; and below, back projection with mirrors at 45° to direction of projection. The apparent origin of both images is point P, thereby minimising distortion. In practice, the mirrors a, for projector A and b, for projector B, may be placed one above the other.

the projectors are arranged with the lens axes parallel to the screen, the mirrors placed at 45° to this direction can be so positioned to give a much closer apparent projection position, thereby reducing the distortion somewhat.

Stereoscopic projection

Whereas a treatment of the subject of stereoscopic photography and projection is beyond the scope of this book, the subject warrants inclusion. The three dimensional appearance of projected stereoscopic pairs of transparencies is so appealing that one is often tempted to give up 'ordinary' photography in favour of this more realistic medium! Two facts must be pointed out, however. Firstly, in order to ensure that each projected image reaches only the eye for which it is intended, the projected images must be polarised (mutually at right angles to each other). Also, audience must wear special polarising spectacles. Secondly, ordinary matt white and beaded screens depolarise the projected images, so that special metallic or aluminised screens must be used. It all sounds like a lot of trouble, but I think it is worthwhile!

Multi screen

The use of two or more screens for slide-tape programmes is gaining popularity. At the moment, most of such programmes are produced professionally, doubtless the increased cost and effort tends to deter most amateurs.

For panoramic views, two or even three screens side by side provide an atmosphere that cannot be matched by the single screen. Also, a most spectacular effect is produced when the entire field of view is filled with a multitude of constantly varying images. Having said these things, I must add some reservations about this medium. There is real danger that the message of a sequence or programme becoming the medium itself, that is, the medium is being used for its own sake, and nothing more. Yes, it is spectacular. Yes, it does impress the uninitiated. But does it convey any information or entertain any better than a well produced single screen presentation? The latter takes quite a lot of beating. Furthermore, when several different image areas are in view, it is all too easy to fall into the trap of permitting one image area to compete with another for attention. Personally, I find it somewhat annoying to attempt to watch changing images in one part of a screen while periodically having to divert my attention to other areas for fear of missing something there. However, as long as a multiscreen presentation has something to say, and as long as it is produced in such a way that the audience can assimilate what is being presented, without distraction, then this medium has much to commend it. In any case, it is with us, and time will tell whether it is just a gimmick or a new and valuable medium.

What about equipment for multiscreen? Obviously, two projectors are required for each screen in use, if fading is needed on all of them. The appropriate control equipment is essential. (Have you ever tried to simultaneously operate four projectors manually?) As far as photography for multiscreen is concerned are special techniques required? Obviously for panoramic views, you must take special care to ensure a level, continuous horizon across the screens. What of other types of images? Is cross-fading so important? Must image brightnesses match on each screen? Must you use all the screens simultaneously or is it acceptable to leave some areas blank? All these questions, and many, many more need answering. But that is the subject for a separate book in its own right!

Projector Lamps

Two different types of lamps are employed for slide projectors,

i) the tungsten filament; rapidly losing favour due to loss, by evaporation, of tungsten which condenses as a black film, on the cooler glass walls thereby reducing the light output.

ii) The tungsten halogen (sometimes called QI – "quartz-iodine") in which iodine is introduced to react with the evaporated tungsten. This does not deposit on the walls of the lamp due to its high working temperature, over 250°C, (hence the use of quartz rather than glass) but condenses back on the filament as tungsten, liberating the iodine to repeat the cycle. This results in a longer life of the lamp.

Handling

The quartz bulb should not come into contact with the fingers as the sodium compounds present in perspiration react with the quartz, at the high working temperature, and thereby permanently impair its optical properties.

Alignment in the projector

Incorrect alignment of the lamp within the projector can not only give uneven screen illumination but also increase any tendency to blackening as well as shortening the life expectancy. Three simple methods are available for checking alignment (i) looking through the lens at the (unlit) filament. (ii) Using a slide of mounted aluminium foil with a pinhole in the centre in the slide gate and using a translucent lens cover with the lamp switched on. (iii) Using a magnifying glass in front of the projector lens to throw an image of the filament onto the screen. In each case the filament, and its mirror-reflected image should be equally sharp but *out* of register ie. displaced.

Care of lamps. Filaments are extremely delicate when operating, hence any movemnt (eg. while aligning projectors) should be minimised and done with great care. (Faulty projector fans can cause vibration.) Decreasing the applied voltage to a lamp by 5% increases working life to about double, whilst only reducing the light output by about 20%. However the reverse is also true so that there is every advantage in using a lamp economy switch, if provided. (It is noteworthy that a 5% increase in voltage on 24v is only 1.2v and the mains voltage can vary by 5%! Have you ever checked the mains voltage selectors on your projector(s)?)

Other optical parts. To enjoy the best performance of the projector lamp, the other optical components of the projector should be clean, this includes mirrors, condenser lenses and projection lenses.

Focusing and Register Slides

The figure opposite can be copied to fill a standard 35mm frame. Two such copies should be made.

The simplest method is to copy onto monochrome film (underexposing and overdeveloping will help to give better contrast). The negatives themselves can be used but they must be mounted with great precision.

Alternatively, if colour reversal film is preferred there can be advantage in making one copy through a deep red filter and the other through a deep green one. The superimposed projected images should give an even yellow colour over the screen. However, these deeply coloured slides are not so convenient for applications 3 and 4 below.

The slides can be used for the following:

i) *Focusing.* The slides facilitate the focusing of projectors.

ii) *Projector alignment.* The slides can be invaluable for checking that the projectors are giving images of identical size and also for the registration of such images.

iii) *Slide gate register.* One of the slides can be used to check whether a projector gate will give consistent image register on the screen. Arrange the projector to give an image at a distance of 10–12 feet. Fasten a piece of graph paper or a piece of paper with vertical lines ruled at about $\frac{1}{8}$ inch (3 mm) apart on the screen in such a position that the centre one of the ruled lines coincides with the projected image of the central vertical line. Reverse the slide magazine, then advance it and check whether the projected central line has changed its position on the screen. About 100 such checks are likely to be needed to obtain a significant result (Note. Ensure that the slide is shown after an *advance* of the magazine, the results may not be the same if the direction of travel of the magazine is reversed.)

iv) *Registration in the mount.* The two slides can be used to assist in the construction of a frame for ensuring precision mounting of transparencies (see page 93).

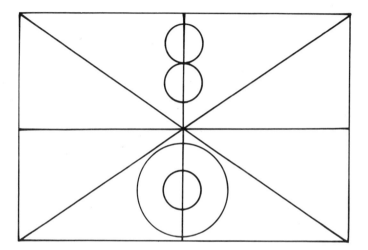

PROJECTED IMAGE SIZES AT VARIOUS DISTANCES

Focal Length in / mm	10 ft (3.1 m)	12 ft (3.7 m)	15 ft (4.6 m)	20 ft (6.1 m)	25 ft (7.6 m)	30 ft (9.1 m)	35 ft (10.7 m)	40 ft (12.2 m)	45 ft (13.7 m)	50 ft (15.2 m)
2.8 / 70	5'0" / 1.5	6'0" / 1.8	7'7" / 2.3	10'2" / 3.1	12'8" / 3.9	15'3" / 4.7	17'10" / 5.4	20'5" / 6.2	23'0" / 7.0	25'6" / 7.8
3.2 / 80	4'4" / 1.3	5'3" / 1.6	6'7" / 2.0	8'10" / 2.7	11'1" / 3.4	13'4" / 4.1	15'7" / 4.8	17'10" / 5.5	20'1" / 6.1	22'4" / 6.8
3.3 / 85	4'1" / 1.3	4'11" / 1.5	6'3" / 1.9	8'4" / 2.5	10'5" / 3.2	12'7" / 3.8	14'8" / 4.5	16'10" / 5.1	18'11" / 5.8	21'0" / 6.4
3.5 / 90	3'10" / 1.2	4'8" / 1.4	5'10" / 1.8	7'10" / 2.4	9'10" / 3.0	11'10" / 3.6	13'10" / 4.2	15'10" / 4.8	17'10" / 5.4	18'2" / 6.1
3.9 / 100	3'6" / 1.1	4'2" / 1.3	5'3" / 1.6	7'1" / 2.2	8'10" / 2.7	10'8" / 3.3	12'10" / 3.8	14'3" / 4.3	16'1" / 4.9	17'10" / 5.5
4.7 / 120	2'10" / 0.9	3'6" / 1.1	4'4" / 1.3	5'10" / 1.8	7'4" / 2.3	8'10" / 2.7	10'4" / 3.2	12'8" / 3.9	13'4" / 4.1	14'10" / 4.5
4.9 / 125	2'9" / 0.8	3'4" / 1.0	4'2" / 1.3	5'7" / 1.7	7'1" / 2.2	8'6" / 2.6	9'11" / 3.0	11'5" / 3.5	12'10" / 3.9	14'3" / 4.4
5.9 / 150			3'6" / 1.1	4'8" / 1.4	5'10" / 1.8	7'1" / 2.2	8'3" / 2.5	9'5" / 2.8	10'8" / 3.3	11'10" / 3.6
7.1 / 180				3'10" / 1.2	4'10" / 1.5	5'10" / 1.8	6'10" / 2.1	7'10" / 2.4	8'10" / 2.7	9'10" / 3.0
7.9 / 200				3'6" / 1.1	4'4" / 1.3	5'3" / 1.6	6'5" / 2.0	7'1" / 2.2	8'0" / 2.4	8'10" / 2.7
9.8 / 250					3'6" / 1.1	4'2" / 1.3	4'11" / 1.5	5'8" / 1.7	6'4" / 1.9	7'1" / 2.2

Projection Distance

APPENDIX 4

DEVICES FOR TWO PROJECTOR CROSS-FADING

Manufacturer Name	Model	Method of fading	Automatic Playback	Rates of Fade	Compatibility	Projectors	Comments	Distributors
Animatic	Convar Mk. 4	fm	Yes	I.V.	With other Mk. 4 models	(1) (7)		Audio Visual Equipment Ltd.
	Convar Mk. 4-B	fm	Yes	I.V.	With other Mk. 4B models	(1) (7)		E. Leitz Inc.
	Mini Convar	fm	No	I.V.	Not applicable	(1) (7)		Wild Pty Ltd.
A.V.E. Diafade	5000 Executive	fm	Yes	I.V.	Not stated	(1) (2)	CAS, M	Diafade
	Diafade	d	Yes	I.V.	Yes with other Diafades and Diafade 4	(1)	IC	
Electrosonic	Q slide ES 69	fm	Yes	I.V.	Yes. Also with ES 3069 and ES 3669	Model A (1); B (2); C (3); D (8)	Z.F.	Electrosonic Ltd.
	Showslide ES 3069	fm	Yes	I.V.	Yes. Also with ES 69 and ES 3669	(1) (2) (3)	CAS, M	
	Showtape ES 3669	fm	Yes	I.V.	Yes. Also with ES 69 and ES 3069	(1) (2) (3)	CAS, M, F microphone/line input mixing facility	Convoy Electrosonic
	Showpulse ES 3006	p	Yes	B	Yes and with many other pulse units.	(1) (2) (3)		Electrosonic Systems Ltd.
	ES 3601	p	Yes	S.T.I.	Yes	(1)	CAS	
	ES 3001	p	Yes	S.T.	Yes	(1)	Z, F, XX	
	ES 3002	p	Yes	T	Yes	(1)	Z, F, XX	
	ES 3004	p	Yes	S.G.	Yes	(2)	XX	
Imatronic	Digital 1000	d	No	I.V.	Not applicable	(4)	IC, F, Q	Imatronic
	Digital 1000T	d	No	I.V.	Not applicable	(4)		
	Digital 2500	d	Yes	I.V.	Not applicable	(4) (5)		
GAF	Manual Dissolve Unit	—	No	I.V.	Not applicable	(9)		GAF
Kodak	Variable Dissolve remote control	—	No	I.V.	Not applicable	(1)		Kodak
Prestinox	FE – M2	m	No	I.V.	Not applicable	(6)	Q	Aico and Prestinox
Purlock	Duofade	fm	Yes	I.V.	Not stated	any	IC	Purlock
Simda	ED 3000	fm	Yes	I.V.	Not stated	(1) (2)	CAS (playback only)	Plemaglen
	15003	fm	Yes	I.V.	Not stated	(1)		

Method of fading

fm = electronic frequency modulation
p = electronic pulse
d = electronic digital
m = mechanical

Projectors

The projectors specified will connect directly with the faders stated. Contact manufacturers or distributors for information on modifying other projectors.
1) Kodak Carousel SAV 2000
2) Eastman Kodak Ektagraphic
3) Leitz Pradovit
4) Any projector but must be modified by Imatronic
5) Connecting plugs supplied for Agfa 250 AV and GAF 203 AV
6) Prestinox 624 AF, 624 AFT, 680 FE, Diatechnic AV*
7) Modification by manufacturer or kit available for home modification
8) Rollei P11 Universal projectors
9) GAF 131 AV and GAF 757 ZS and any projector with AV facilities and 24v, 150–250w halogen lamp

*NOTE. Projector must have inbuilt triacs for fading and selector switch for fading/normal use.

Footnote

Snap changes are virtually instantaneous being achieved by a mechanical shutter. Involves modification of the projector

(1 or others) by manufacturer of fader.

T   Permits choice of any two rates from 2.5, 4, 6, 9, and 12 secs. in any one sequence.

I   Superimposition of two slides possible in place of snap change. Can also be used to stop programme at pre-determined points.

Z   Automatic zeroing of magazine possible.

F   Facility for "flashing" extinguished projector.

B   Fast dissolve and reverse. Adjustable rate of fade 3–12 secs.

Q   Separate slide-tape synchroniser can be used to provide visual cue for projectionist (on hand control).

G   Choice of 5 rates of fade 2.5, 4, 6, 9 and 12 seconds.

CAS   Includes cassette deck, amplifier and speaker.

M   Includes microphone.

IC   Permits independent control for each projector.

XX   Designed to serve also as a part of existing multiscreen systems.

Apparatus and Instrument Co. Ltd.,
Aico House,
Alexandra Road,
Hounslow,
Greater London Borough of Hounslow,
England TW3 1JT.

Audio Visual Equipment Ltd.,
73 Surbiton Road,
Kingston,
Surrey,
England KT1 2HG.

Diafade Limited,
116 The Maltings,
Stanstead Abbolts,
Ware,
Hertfordshire,
England SG12 8HG

Convoy Electrosonic,
4 Dowling Street,
Woolloomooloo 2011,
Sydney,
Australia.

Electrosonic Ltd.,
815 Woolwich Road,
London SE7 8LT, England.

Electrosonic Systems Inc.,
4575 West 77 Street,
Minneapolis,
Minnesota 55435, U.S.A.

Imatronic Ltd.,
Dorian House,
Rose Street,
Wokingham,
Berkshire, England RG11 1XU.

GAF (Great Britain) Ltd.,
Photo Products Division,
P.O. Box 70,
Blackthorne Road,
Colnbrook,
Slough, SL3 0AR, England.

Kodak.

E. Leitz Inc.,
Rockleigh,
New Jersey, 07647, U.S.A.

Plemaglen Audio Visual,
36 Southwell Road,
London SE5 9PG, England.

Prestinox,
Route de Tremblay,
93420 Villepinte,
France.

Purlock Duofade,
Obtainable from:–
Sir George F. Pollock, Bt., F.R.P.S.,
Netherwood,
Stones Lane,
Westcott,
Dorking, Surrey.

Wild (Australia) Pty Ltd.,
45 Epping Road,
North Ryde,
N.S.W. 2113,
P.O. Box 21,
Australia.

Control of More than Two Projectors

Economics, the complexity of the sequence and the frequency with which such sequences will need to be shown will dictate, to a large extent, whether the use of three or more projectors is justified. The following suggestions are listed in order of increasing complexity and increasing demand on the facilities offered by the equipment used (and excludes the obvious method of manual control by a separate projectionist on each pair of projectors).

Three projectors

(i) Automatic control of projectors 1 and 2 and manual control of projector 3.

(ii) Automatic control of projectors 1 and 2 and control of projector 3 by synchroniser (external type preferably, internal pause type would involve a pause in the soundtrack on stereo recorders. If a third channel is available for the automatic control then the pause type synchroniser can be used with a stereo sound or pulse type with mono sound).

(iii) Diafade 4 controls 2, 3 or 4 projectors giving infinitely variable rates of fade on each independently. Impulses can be recorded on single track only. Compatible with Diafade (Appendix 4).

Four projectors

(i) Automatic control of projectors 1 and 2 and manual control of 3 and 4.

(ii) Automatic control of projectors 1 and 2 and automatic

control of 3 and 4 (this involves two fading devices and an extra track on the tape ie. a minimum of three channels if mono sound is used or 4 channels for stereo sound).
(iii) Diafade 4 (see above).

More than four projectors

Obviously with a four channel tape deck, up to six projectors can be controlled via three faders. Commercially available devices for projector control for a single screen or multi-screen include
(i) Simda ED 3300, controls up to six projectors, in pairs, permitting infinitely variable rates of fade.
(ii) Electrosonic Autopresent units (ES 3681/2/3) control up to eight projectors via faders ES 3001 and ES 3002 (six projectors with ES 3004).
(iii) Electrosonic ES 3003 controls individual projectors giving eight rates of fade. Up to fifty six projectors can be accommodated by use of a central controller.
(iv) Kodak Carousel SAV 2000 Programme control. The unit can control up to ten single projectors or ten pairs of projectors if linked with the Quick Change Control or Variable Dissolve Control (not listed in Appendix 4). The programme can be recorded and up to 15 different functions can be programmed.

Cassette Recorders with 3 and 4 Channels

When stereo sound and automatic projection is required at least three channels of a tape deck are needed. For automatic control of four projectors and stereo sound then four channels will be required. There are a few four channel open reel decks available (your local hi-fi dealer should be able to help you) but these tend to be inconvenient for transport. However, at the time of writing, there is one make of a series of high quality cassette machines available which has gained great popularity in this medium. These are the NEAL machines.

Three channel decks

102 AV Stereo cassette deck with Dolby noise reduction on both channels. Extra channel for projector control purposes. Independent "record" or "playback" on all three channels.
103 AV As for 102 AV but with three stereo input mixing or six mono input mixing facilities.

Four channel decks

140 Four channel cassette deck with Dolby noise reduction on all four channels, or without Dolby on specified channels.

SUPPLIERS

UK. North East Audio Ltd.,
    Simonside Works,
    South Shields,

Tyne and Wear,
NE34 9NX,
England.

Australia Concept Audio Pty. Ltd.
13 Richmond Road,
Narrabeen,
N.S.W. 2101

Canada H. Roy Gray Ltd.,
14 Laidlow Boulevard,
Markham,
Ontario.

USA. Audiosource,
1185 Chess Drive,
Foster City,
California 94404

Copyright – Sound Recording

In Great Britain, the Commonwealth countries and the United States, the law is quite clear; it is *illegal* to record from disc, tape, radio, television, and in many cases, live performances, for any purposes whatsoever, without the consent of the holder of the copyright. The object of this is equally clear; to ensure that the originators, arrangers and performers of music and other sound effects obtain their just desserts from their works and to prevent 'piracy' by others.

The interpretation of the law, however, seems less clear. There are those who claim that it is illegal to record for *personal gain* and hence, recording for personal pleasure only (eg. "pop" music for parties) is permitted. Furthermore, some maintain, the implementation of the law on copyright is virtually impossible, and therefore they are safe to record whatever they wish. Recording for slide-tape sequences is believed, by many, to be permitted as long as no fee is charged for attending the performance of such sequences. None of these arguments are necessarily tenable in law, and it would be advisable not to become the defendant in any case which may arise.

The answer to this problem is simple; obtain a license to record and use copyright material, by contacting one of the bodies below (and if my experience is typical, you will find them extremely helpful!).

Mechanical Copyright Protection Society Ltd.,
Elgar House,
380 Streatham High Road,
London SW16 6HR.

National Music Publishers Association Inc.,
110 East 59 Street,
New York,
NY 10022.

Australian Music Publishers Association Ltd.,
P.O. Box Q,
123 Queen Victoria Buildings,
Sydney,
N.S.W. 2000

Copyright – Photography

The law on copyright varies from country to country although Britain and the Commonwealth, the USA and many countries of Europe and the South American continent are co-signatories to agreements affording the copyright owner of a photograph the same protection as exists in his own country or state. (The law also applies to any other artistic, literary, dramatic or musical work.) The word "photograph" relates to an "original artistic work" produced by photographic or similar processes. Hence a specific transparency comes under the laws of copyright but not the *idea* on which the photograph was based. The copyright lasts, in Britain, for fifty years after the photograph was made, and for the same period of time after its first publication.

The ownership of the copyright of a photograph rests with the owner of the material on which the photograph was originally taken, unless the photograph was commissioned by another party. In the latter case the copyright is owned by the party commissioning the photograph even though the photographer may be the owner of the photograph itself. In the case of photographs commissioned by the proprietor of a newspaper or periodical, the photographer retains the copyright but the proprietor has the first right to publish. The ownership of copyright can be sold, or given away to another, but, for legal purposes, written evidence of such transfer of ownership is needed. Legal action can be taken, by the copyright owner, against any party using photographs without permission.

Slide-tape Synchronisers

EXTERNAL

Philips N6400	Placed at the side of open reel decks. Impulses are recorded on track 4, permitting tracks 1 and 3 to be used for sound	Philips Electrical Ltd. City House, 420/430 London Rd., Croydon, ENGLAND CR9 3QR.

INTERNAL
Pulse Type

one channel needed for impulses

(a) Helikon SS100	Cassette or open reel	Hammonds of Watford, 47 Queens Rd., Watford, Herts ENGLAND WD1 2LW
(b) Philips N6401	Designed principally for N2209 & N2229 cassette recorders.	Philips Electrical Ltd. 420/430 London Rd., Croydon ENGLAND CR9 3QR

INTERNAL
Pause type

(c) Highgate-Dufay	Cassette or open reel Mono or stereo. Slide-change mechanism actuated by pause in sound, duration adjustable 1–9 seconds. Any other pause of shorter duration than the chosen one is without effect.	Highgate-Dufay Ltd. 38 Jamestown Rd., London, ENGLAND NW1 7EJ

(d) Redicoupler

Rediffusion
   Reditronics Ltd.
La Pouquelaye,
St. Helier,
Jersey.
Channel Islands.

A Circuit for a "Buzz-box"

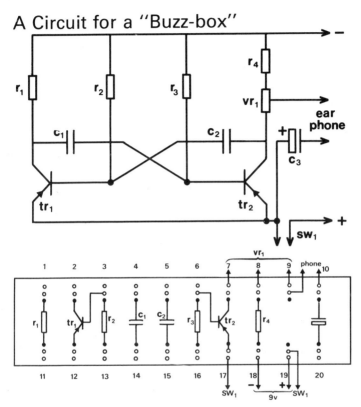

'Buzz box'. The circuit can be constructed on Vero Strip or Vero Board wide enough to take the components and with at least 2 holes spare at each side. Each copper strip should be divided in the centre by careful use of a twist drill. If the signal generator does not work even after checking the wiring and battery (i) Disconnect earphone and short across the switch contacts. (ii) Use a meter across contacts 1 (+ve) and 11 (-ve). If meter gives no reading tr1 is malfunctioning. (iii) Use meter across contacts 7 (+ve) and 17 (-ve) to check function of tr2. Components. r1 — 15 KΩ, r2 — 33 KΩ, r3 — 33 Ω, r4 — 4.7 KΩ (all resistors 1/3 watt). c1 — 0.1 μF, c2 — 0.1 μF, c3 — 0.02 μF electrolytic. tr1 and tr2, any small signal, low current PNP transistors e.g. 2N 4059. vr1 — 10 KΩ, earphone — 1000Ω (preferably), board — Vero Strip board, battery — 9v (PP3). Constructions. To complete the circuit, join contact strips — 1, 2 and 4; 5 and 7; 3 and 15; 6 and 14; 11, 13, 16 and 18; 12, 17 and 20. (The author is indebted to Mr. Colin Matthews for the design.)

191

Suggested Checklist

1. Projector 1
2. Projector 2
3. Leads for projectors
4. Remote control leads for projectors
5. Cross fader (with leads if needed)
6. Tape deck and power lead
7. Amplifier and power lead
8. Leads from 6 to 7
9. Speakers
10. Speaker leads (and extensions)
11. Synchroniser
12. Buzz box or alternative
13. Ear piece for 12
14. Headphones
15. Screen
16. Support for projectors
17. Spare projector lenses
18. Microphone
19. Universal plug(s)
20. Multiple sockets
21. Register slides
22. Magazine of slides
23. Tapes or cassettes
24. Taped background music
25. Spare bulbs
26. Spare fuses
27. Cue sheet
28. Stop clock
29. Adhesive tape

Screens and Viewing Conditions

Good quality transparencies and projection equipment deserve good quality screens. Similarly your audience deserves satisfactory viewing conditions. To a certain extent, the latter are governed by the type of screens in use. The matt white screen is the lesser expensive and is generally available in the larger sizes whereas the glass beaded screen gives a brighter image, albeit over a narrower angle. This is by far the better screen for a long, narrow room. The best effect is obtainable within about 15° each side of a line perpendicular to the centre of the screen. For the audience seated between 15° and 25° each side of this line, the glass beaded and matt white screen perform equally well, whereas beyond 30° a matt screen is a "must" although its performance decreases rapidly as the viewing angle increases further. In addition, for comfortable viewing (and assuming the screen is square, to accommodate horizontal and vertical formats equally well) no member of the audience should be nearer than twice the height of the screen (the room height is often the limiting factor for screen sizes) nor further away than eight times the screen height.

Index

Advantages of using two projectors 29–31, 126
Alignment of projectors, 38–40 174
Amplifiers 77–79
Anti-Newton's ring mounts 92
Atmosphere, recording 138
Automatic focusing 35
  and projector fading 91
Automatic projection 18, 64, 87
  advantages and disadvantages of 52
  with three or more projectors 182
Automatic slide changing 14

Back projection 168
Bas relief, by projection 132
Blank slides 25, 97
Blanking plug 58
Buzz box 86, 191

Cassette recorder 67–69
Changing by cross-fading focus 127
  monochrome and colour 130
  lighting conditions 130
  subjects 129
  time of day 129
Changing focus 125
Channel selectors, need for separate 69
Checklist 152, 192

Colours, mixing 167
Commentary
  incorporating 142
  writing 143
Compatability
  of fader with fader 60
  of tape deck with fader 60
Copyright
  photography 188
  recording 117, 137, 140, 186
Cross-fading
  electrical methods 52, 55
  mechanical methods 49, 52
  methods of 49
Cuesheet 84, 118

Deterioration of slides 98
Dolby noise reduction 66, 68
Dual projectors 61
Duration
  of individual slides 24, 157
  of sequences 155

Echo effects 74
Editing
  slides 103
  sounds 112
Electrical faders
  advantages and disadvantages 55
  home-made 58
  modification of projectors for 56
External synchronisers 18, 20, 76, 85, 189

Fader compatibility 60
Faders
  electrical 55, 56, 58
  mechanical 49–52, 54
Filters
  colour, for projector alignment 39, 174
  on amplifiers 79, 140
Flip-flop method of image registration 132
Focusing slides 39, 40, 174
Four channel recorders 162, 163
Four projectors, control of 182

Ghost effects 128

Headphones 73, 152
Highlights and shadows, use of 127
Horizontal and vertical formats 106
  methods of changing 134
Human voice, choice for sequences 137

Image distortion
  avoiding keystoning 153, 168
  non-rectangular 37–38
  obscuring effect of 43, 168
  with back projection 168
Image and sound matching 24, 101, 104
Impedance 80
Instant stop key 70

Internal synchronisers 18, 189, 190
  pause type 20, 86
  pulse type 20, 86

Jerk change 135

Keystone effect 36
  avoiding 153
  obscuring 43, 168

Lenses, projection, choice of 44
Lightning effects 128
Light table 23, 96, 103
Line and lith film
  for colour derivatives 132
  for title slides 108

Magazine
  advance, non-alternate 54
  changes 15, 150
  rotary, advantages and disadvantages 16, 150, 151
  straight, advantages and disadvantages 15, 150, 151
  transfer storage 16
Manual projection 18, 64
Matching images and sounds 24, 101, 104
Mechanical faders 49–52
  positioning of 50, 52, 54
Microphones 82
Misregister
  causes 32
  effect of projector gate 33

effect of transparency mount 32
obscuring the effect of 43
temporary 40
Mixers 67, 140
and microphone impedance 82
types of 141
Mixing
colours by projection 167
down 141
facilities on the recorder 73
horizontal and vertical formats 106, 134
Mixing sounds
and commentary 142
down from stereo to mono 141
facilities on the recorder 73
mono to simulate stereo 141
Monitoring 72
control signals 76
on location 138
Mono and stereo sound 63
for location recording 138
Mounting slides 91
effect of parallax 94
for precision in register 93–96
Multi screen techniques 170
Music, choice of 24

NEAL tape recorders 162–163, 184
Newton's rings 92
Non-alternate showing of slides 135

Numbering slides 16, 96

Open reel recorders 67–69

Panning images and sounds 106, 162
'Panpots' on mixers 140
Parallax, effect on precision in mounting 94
Pause control 70, 112
Precision
in positioning of extra tape head 66, 71
in slide mounting 32, 93
Presentation effects with two projectors 126–136
Projection
automatic 18, 52, 64, 87
with three or more projectors 182
manual 18
Projector bulbs 42, 172
Projectors
alignment of 38–40
positioning of 46
slides for alignment of 174
Projector lens
choice of 44
choice related to fading technique 45
Projectors
choice of, for register 34–36
matching 44
Projector support 40, 47
Punch-pin, method of image registration 130
Rate of slide changing 24

Recorder, choice of 18, 63–77
Recording 111
  atmosphere 138
  on location 137
Register
  flip-flop method 132
  images on screen 153, 174
  importance of 42
  methods of maintaining
    with derivative slides
    130
  punch-pin methods for 130
  slide mounts for precision
    in 93, 175
Registration
  of derivative photographs
    130–132
  of images on screen 153,
    174
  slide mounts 93
  slides for 174
Remote control lead with
  non-automatic projec-
  tion 14

Screens 196
Scriptwriting 143
Slide changing
  methods with different
    projectors 15
  timing of 50, 83–87, 117
Slide mounts
  choice of 92
  correct use of 93
  precision register 32, 93
Slides
  arranging in order of projection 105

Slides
  blank 25, 98
  deterioration of 99
  editing 103
  horizontal and vertical
    formats 106
  mounting 91
  spotting and numbering
    16, 96
  storage 90, 99
  title 107
Snap changes
  effect of fading device 49,
    56
  for changing formats 135
  general use of 135
Sound accompaniment
  matching images 24, 101,
    104
  mono or stereo 63
  value of 16, 62
Sound and image matching
  24, 101, 104
Sound effects records 140
Speakers 77, 79
  impedance 80
Split screen 164
Spotting slides 16, 96
Square format, use in cross-
  fading 134
Stereoscopic projection 170
Stereo sound
  and automatic projection
    64, 161
  mixing down to mono 141
Stopclock and stopwatch,
  use of 25, 84
Storage of transparencies

mounted 90
unmounted 89
Synchronisers
external 18, 20, 76, 85, 189
internal pause and pulse types 18, 20, 86, 189, 190
use of 26, 85, 86, 119
with three projectors 166, 182
Synthetic colour effects 132

Tape loop 139
Tape recorder
channel selector switches 69
choice of 63–77
compatibility with fader 76
conversion for stereo and auto playback 66
echo effects 74
four channel (quad-raphonic) 64, 162
open reel or cassette 67–69
pause control 70, 111
three channel 66, 164
transportability 66
Tape speed 68, 112, 163
TEAC recorder 163
Three projectors
on one screen 165

control of 182
Time lapse effects 128
Timing slide changes 50, 83–87, 117
Title slides 107–110
Transfer storage magazines 16
Twinkle effects 60, 135
Two projectors, advantage of using 29–31, 126

Vertical and horizontal formats 106
methods of changing satisfactorily 134
Viewing conditions 193
Vignetting
with mechanical faders 45, 50
obscuring the effect of 52, 54
Voice
choice for sequence 138
-over, incorporation with sound 72, 142
VU meters 74

Zoom lenses for projection 45, 52